Also available in the Bloomsbury Sigma series:

GENUINE FAKES

How Phony Things Teach Us About Real Stuff

Lydia Pyne

BLOOMSBURY SIGMA
LONDON · OXFORD · NEW YORK · NEW DELHI · SYDNEY

BLOOMSBURY SIGMA
Bloomsbury Publishing Plc
50 Bedford Square, London, WC1B 3DP, UK

BLOOMSBURY, BLOOMSBURY SIGMA and the Bloomsbury
Sigma logo are trademarks of Bloomsbury Publishing Plc

First published in the United Kingdom in 2019

Photo credits (t = top, b = bottom, l = left, r = right, c = centre)
Colour section: P. 1: Image courtesy of Kaye et al 2015 / Creative Commons
license (t); Photos by L. Pyne / Oxford University Museum of Natural History
(bl, br). P. 2: © Classic Image / Alamy Stock Photo (t); © age footstock / Alamy
Stock Photo (b). p. 3: © 1988.125 / Spanish Forger. Betrothal of St. Ursula.
[19–] © The Morgan Library & Museum. 1988.125. Gift of Martin Cooper. (t);
© Harry Ransom Center (bl, br). P. 4: © Science History Institute / Creative
Commons license (t); © Museum of Innovation and Science (bl, br). P. 5:
© New York Public Library (t); © Science History Institute / Creative Commons
license (b). P. 6: Photo by L. Pyne / University of British Columbia (t); Hanging
model of the Blue Whale in the Irma and Paul Milstein Family Hall of Ocean
Life © American Museum of Natural History (b). P. 7: Screenshot by L. Pyne
© Explore.org / Alaska Department of Fish and Game (t) © Daniel J. Cox /
Getty Images (c); Photo by Carole Fritz © Caverne du Pont d'Arc (b). P. 8: Photo
by Christian Tran © Caverne du Pont d'Arc (t); © PA Images (b).

A catalogue record for this book is available from the British Library

Library of Congress Cataloguing-in-Publication data has been applied for

ISBN: HB: 978-1-4729-6182-2; TPB: 978-1-4729-6183-9;
eBook: 978-1-4729-6181-5

2 4 6 8 10 9 7 5 3 1

Typeset by Deanta Global Publishing Services, Chennai, India
Printed and bound in Great Britain by CPI Group (UK) Ltd, Croydon CR0 4YY

Bloomsbury Sigma, Book Forty-seven

To find out more about our authors and books visit www.bloomsbury.com
and sign up for our newsletters

For Stan

Contents

Warhols Without Warhol

No one wants to be bamboozled by a fake, but everyone loves hearing about those who are.

Frauds, forgeries and fakes all make for fantastic stories and have for millennia. In ancient Rome, for example, shrewd art collectors were wary of cheap knock-offs of valuable Greek vases and sculptures. The famous philosopher Cicero was thought to have had rather discerning taste, collecting only the most authentic of Greek art; statesmen like the general Sulla and emperor Nero, art-savvy Roman patricians sniffed with disdain, did not. The Middle Ages saw a rise in the dubious – yet lucrative – economy of selling 'genuine' religious relics to gullible wayfarers on religious pilgrimages. ('As for bones of St. Denis,' Mark Twain quipped after touring medieval reliquaries of Europe centuries later, in the 1860s, 'I feel certain we have seen enough of them to duplicate him, if necessary.') Talented forgers have hoodwinked collectors for centuries, lining their pockets and revelling in having pulled the wool over the world's eyes. Running the gambit from clever hoaxes to embarrassing swindles, the history of fakery is certainly never dull.

But fakery isn't a phenomenon that only inhabits the worlds of art and antiquities. Nothing, it would seem, is safe from a faker's clutches, and all manner of things – from paintings to fossils, rare books to flavourings, gems to artefacts in museums, and even nature itself – have at one time or another been faked, and faked spectacularly.

It's easy to treat 'real' and 'fake' as discrete, distinct categories, because finding examples of each appears to be rather straightforward. Designer handbags sold in Saks?

Real. Knock-off purses hawked from a corner, where Gucci is spelled with one 'c'? Fake. The *Mona Lisa* in the Louvre? Real. Da Vincis you can buy on eBay? Fake. Living history museums? Real. Renaissance Faires? Fun, but fake. A raclette wheel? A block of Wisconsin cheddar? Real. Cheez Whiz? Fake. Definitely fake.

But what happens when it becomes trickier to sort out what is real and what is not? Do the same things that make something real also make it authentic? What are we to think when a fake becomes even more famous than its original? Could a fake object meet our expectations for authenticity better and more directly than the genuine one ever could? Or could artificial objects be more desirable – more ethical perhaps – than natural ones? How do older standards for authenticity translate into the twenty-first century?

It turns out that the world is full of things that defy a neat, superficial categorisation – it's full of in-between objects that are real and not-real at the same time. They're what we might call 'genuine fakes'. Sometimes we think that they're authentic, sometimes not. They're provocative and fascinating and challenging. And they're everywhere.

★ ★ ★

The American artist Andy Warhol died on 22 February 1987. But the small technicality of his death doesn't necessarily mean that there aren't new Warhol paintings to be made, sold and collected.

In 2010, the artist Paul Stephenson came across 10 original Warhol acetates from the mid-twentieth century. (Acetates are the 'negatives' used in silk screening.) Stephenson purchased them, although at the time he wasn't sure what exactly he would use them for. The acetates included several iconic Warhol motifs – *Jackie*

Kennedy, Mao and even Warhol's own self-portrait – and were quickly authenticated by the Andy Warhol Museum in Pittsburgh, by Alexander Heinrici, Warhol's own master printer, and by art expert Rainer Crone. Warhol himself, Stephenson was told, had left the paint on the acetates.

After extensively researching Warhol's painting methods, Stephenson decided to use the acetates and, working with Heinrici, created a new set of prints from the negatives by employing the same methods that Warhol had used to create his originals. As the BBC reported, Stephenson used the 'same silkscreen inks, stretcher bars, canvas, everything the original artist would have used'. To some, Stephenson was essentially creating new Warhols – although Stephenson is quick to say that he doesn't consider his prints to be originals. The series of paintings was called *After Warhol*, a collection Stephenson describes as a 'forced collaboration' because the original artist couldn't possibly be aware of it.

Historically, Warhol had a lot of assistance in making his prints. Not for nothing was Warhol's studio dubbed The Factory, as many assistants and workers did most of the physical work of painting and printing – Warhol simply added the finishing touches. On some occasions, Warhol's assistant and even his mother signed his paintings on his behalf.

Rainer Crone (who passed away in 2016) suggested that these Stephenson-made Warhol prints could be considered authentic, and that at some point in the future they might even be catalogued as such. According to an October 2017 interview with the BBC, when Crone saw Stephenson's prints he proclaimed, 'paintings made with these film positives under described circumstances and executed posthumously by professionals (scholars as well as printers) are authentic Andy Warhol paintings'. The Warhol Museum in Pittsburgh pointed out in the same BBC interview that, yes, Stephenson's Warhol prints were in keeping with the spirit of Warhol's work, but that Warhol himself always had some touch to add in every print and that obviously couldn't happen with the *After Warhol* prints; the museum also described the concept of this forced collaboration as 'problematic'.

After Warhol has, needless to say, prompted a flurry of questions about realness, authorship and authenticity, and the prints are perfect examples of genuine fakes. 'If the world-leading Warhol scholar says it's a Warhol, and you do everything in the mechanical process that the original artist did, and the original artist said "I want other people to make my paintings," which he did – what is it?' Stephenson offered to the BBC. 'I don't know the answer to that question.'

Incidentally, in 2011 the Andy Warhol Foundation surprised the art world by dissolving its authentication board and stating that it would no longer subject itself to the ongoing hassle and legal headache of authenticating any piece of art that wasn't already in Warhol's recognised catalogue raisonné. (An estate or foundation's authentication boards serve as the 'official' arbiters over which pieces of art can be certified as those of a particular artist and belong in the collectively agreed upon catalogue of known works and which pieces cannot.) Between 1995 and 2011, the Warhol Foundation had its board examine some 6,000 purported works of Warhol's – some real, some not – and finally folded up shop, due to the inexorable financial toll of legal lawsuits filed by disgruntled collectors. 'One year our legal bill ran up $7 million,' said Joel Wachs, the foundation's director, in a 2015 interview with *Authentication in Art*. 'The cost to defend them became so great, we got tired of giving money to lawyers. We'd rather be giving it to artists.'

As a result of disbanding the board, any future Warhol paintings that go up for auction will do so without the board's appraisal. Pieces that have already been authenticated – that is, accepted as legitimate Warhols and grouped in his catalogue raisonné – have proven to be particularly valuable in the collecting world. For example, Warhol's *Triple Elvis* (1963) sold for $82 million at Christie's in 2014, three years after the authentication board had disbanded. Arguably, the success of the sale leaned heavily on the fact that *Triple Elvis* had already been authenticated.

Works of art that are generally considered rare objects themselves have been made all the more scarce when their authentication is treated as fixed and an artist's catalogue raisonné as non-changing. There are contemporary art experts, like Richard Polsky, who will authenticate Warhol works, but that authentication is independent of the Warhol board. When the BBC asked Polsky about the *After Warhol* prints, he stated, 'I like the fact that he [Stephenson] is honest – he's not claiming Andy made these, he's claiming he made them.' Polsky applauded the 'modest' price attached to the Stephenson prints, but expressed some of the same concerns voiced by the Warhol Museum, 'It sounds like he's trying to extend Warhol's career, so to speak, even though he's dead. There's a charm to that, but it just seems so shallow.'

Since the Warhol Foundation's decision to disband its authentication board, a number of other artists' estates and foundations – including those that represent Jean-Michel Basquiat, Roy Lichtenstein, Keith Haring and Jackson Pollock – have also opted to retire their boards, rather than deal with the legal repercussions of mistakenly authenticating some work of art that later proves to be fraudulent. Within the last decade, scholarly conferences that focus on the authenticity of an artist's work have been cancelled, as even the merest whisper of doubt about a painting could have ramifications for its value. 'In the high-stakes art world,' art journalist Stacy Perman concluded after her 2015 reporting on the disbanding of the Warhol authentication board, 'a fear of lawsuits is putting a muzzle on authenticators.'

★ ★ ★

The 'Could Paul Stephenson's prints be new Warhols?' story is particularly provocative because it encourages viewers to really think about where a 'real' object ends and where a 'fake' (or less than real) object begins.

On the surface, this seems hilariously simplistic. *Of course*, most people would agree, rolling their eyes, if a painting is going to count as a 'Warhol', then Warhol would have had to have painted it himself while he, technically, was alive. (This stands in contrast to work that is simply published posthumously; *Jurassic Park* author, Michael Crichton, for example, has had three sci-fi novels from beyond the grave.) Stephenson's art asks what it takes to make something Real and something Not.

The concern about authenticity – and what to make of Stephenson's 'genuine fakes' – isn't a conundrum that is unique to the art world's authentication boards. The problems, costs and curiosity of authentication spill over into other markets, such as antiquities, rare books and manuscripts, the flavour of food and even fossils. Again, it's easy to treat 'real' and 'fake' as discrete, distinct categories, but more often than not, concerns about whether something is real or not are actually concerns about authenticity – specifically, how authenticity is translated into cultural and financial value. Authenticity shapes how we evaluate the material world around us, as well as how we think about intellectual property, representation and even history. Amid worries about 'fake news' and 'alternative facts', the question of authenticity has taken on particular urgency in the twenty-first century.

The late-twentieth-century American philosopher Denis Dutton offered a distinction about authenticity that feels apropos to this very slippery continuum of Real and Not Real. Dutton suggested that authenticity could be couched as 'nominal' when a thing is correctly attributed to its author (and not a forger), and that 'expressive authenticity' could be conveyed through a work to an audience by alluding to values, feelings and beliefs – what Dutton called 'inherent authority'. In other words, there are many ways in which a work of art – or any object – can and ought to be considered authentic, and its authenticity can change over time and keep pace with history.

If Stephenson's Warhols were made of, say, painted vinyl instead of canvas, it would be clear that, materially, they're not in the spirit of a genuine Warhol. Likewise, if Stephenson had tried to pass off his paintings without the story – the provenance – of how and why he made them, they would no longer be interesting originals, but would cross over into forgeries. Consequently, intent, provenance, material and history all matter if we're to sort out what fakes matter, how they matter and why. Picking through the complicated stories of genuine fakes ensures that questions of authenticity and authorship are asked and answered, and asked again.

★ ★ ★

One of the hardest parts of writing this book was deciding which genuine fakes and their stories to include. For every object discussed here, there are three, or four, or five equally interesting things that I ended up cutting. As the book came to life, friends and colleagues sent me articles and suggestions for quirky, eclectic, bizarre objects that were all brilliant examples of genuine fakes. 'I can't believe this is really a thing!' became a popular refrain from those forwarding articles to my inbox.

What to include? And, much harder, what to cut? In the end, I chose objects to write about that piqued questions about authenticity and that I felt didn't have straightforward answers. What happens when forged paintings, like those of the Spanish Forger, become collectable in their own right? Should we still think of them as fakes? But authentic fakes? How can faux fossils from 1725, created as a prank, help us understand what people thought about the natural world almost three centuries ago? Can artefacts like the ancient Maya Grolier Codex ever really be accepted as authentic when their discoveries and provenances are so problematically unreliable?

Pushing these questions even further: as the twenty-first century hones technology that is better at copying objects from nature, these replicas take on their own questions of ethics. If laboratory-grown diamonds are, on a material level, identical to natural diamonds, what's to separate the two gems? Consumer pressure? Is it possible, then, for the 'fake' to be more ethical than its natural counterpart? The same goes for synthetic flavours – what parts of nature can be authentically replicated, and what parts can't? When does a model – or replica or copy – sufficiently stand in for the 'real thing' in museum collections and at tourist sites, and when does it come across as an obviously faux? With all of the ways to watch the wilderness – livestream, documentaries and so on – what is the most real, most authentic way to see the natural world if you can't visit it in person? And what are the trade-offs with each of these alternatives?

Although the topic of 'fake news' is overwhelmingly omnipresent in today's media, I have not included it here; I think historians of propaganda can better offer the subject the context and nuance that it deserves. (I would heartily recommend Kevin Young's *Bunk: The Rise of Hoaxes,*

Humbug, Plagiarists, Phonies, Post-Facts, and Fake News.) The genuine fakes described in this book are fundamentally rooted in the material world – they are tangible, physical things that have been made, unmade and remade any number of ways throughout history. They are things that have challenged how I think about authenticity.

★ ★ ★

To put this all another way: when Mark Twain was travelling through Europe and the Holy Land in 1867 he encountered a plethora of ways in which depicted history – particularly history that was presented to tourists – was less than genuine. When he wrote up his experiences in *The Innocents Abroad*, he was pretty darn sure that there were enough 'authentic pieces of St. Denis to assemble the saint's skeleton multiple times over'. The idea that parts of St Denis could be in so many places at once simply strained credibility. 'Isn't this relic matter a little overdone? We find a piece of the true cross in every old church we go into, and some of the nails that held it together. I would not like to be positive, but I think we have seen as much as a keg of these nails,' Mark Twain remarked while considering the European relics that he and his travelling companions encountered during their tour. 'Then there is the crown of thorns; they have part of one in Sainte Chapelle, in Paris, and part of one, also, in Notre Dame.'

There wasn't any way that all of those bones could be 'real' bones of St Denis, or all of the nails could have come from the Cross. But Twain also hints at the desire for authenticity that these relics inspired – and how, after centuries of being less than genuine, the relics offered a sort of realness in the fakery. What mattered was whether people wanted the bones to be real. Twain suggests that this desire for realness works like a cultural placebo. If the bones are real enough to resonate with audiences, then they're good enough to be considered the genuine thing. 'We did not feel desire to disbelieve these statements,' Twain mused. 'Yet we could not feel certain that they were correct.'

The world is full of genuine fakes, and the line between real and not isn't a sharply drawn boundary. Genuine fakes live along a continuum of authenticity – a gradient that unfolds narrative after narrative, story after story. Fakery has an uncanny ability to unsettle the cultural status quo as it challenges how things are made Real.

The ancient Roman satirist Petronius reminds us that, 'The world wishes to be deceived. So let it be deceived.' In a twist of beautiful historical irony, it is unclear whether or not Petronius actually ever, technically, uttered that particular adage. But its sentiment stands.

CHAPTER ONE

This Solemn Mockery

On Wednesday, 23 May 2012, the British auction house Bonhams began accepting bids for items from one of the most impressive collections of forgeries amassed during the twentieth century. There were phony letters attributed to the poet Percy Bysshe Shelley and the novelist George Eliot. A few faux medieval panels and paintings. Some fake Shakespearean ephemera. Ardent enthusiasts of all famous historical things fake, welcome to the sale of the celebrated Stuart B. Schimmel Forgery Collection!

During his life, Stuart B. Schimmel was a respected collector of rare books, manuscripts, engravings and historical printing technologies, following his successful career in business and accounting. 'He brought verve and élan to the grand collection he amassed,' John Neal Hoover lauded in his 2013 memorial, noting that Schimmel was 'a true adventurer in the pursuit of books and the ways they can order one's life'. In addition to building a traditional art portfolio with works by recognised masters, Schimmel spent half a century collecting superb forgeries, frauds and fakes.

Schimmel got his start collecting forgeries when he offered to buy a Lord Byron signature from his friend, Colonel Drake, who had unwittingly purchased a very good Byron forgery. (Schimmel graciously agreed to sell the signature back to Colonel Drake if, at some point in the future, it proved genuine.) From there, Schimmel went on to collect only the very best fakes, with items boasting amazing provenances and stories, rarities that commanded impressive prices. ('Provenience' to refer to the position of an artifact's discovery, like at an archaeological site. 'Provenance' refers to the cultural and custodial history of an artifact or object once it has entered a museum or collector's context.) He collected the ersatz with enthusiasm as well as a discerning eye. All of the fakes in his collection, you see, had been authenticated.

'The motivation for forgery is always complex, whether done for gain or fame, to prove a point or in the belief that something should exist even if neglected by its purported author,' renowned book collector Nicolas Barker suggested in the introduction to the 2012 Bonhams catalogue that listed the Schimmel lots. 'There is no such difficulty with Stuart's collection. All his are genuine forgeries. He has had a lot of pleasure out of collecting them, and the books about them. It is now the turn of others to enjoy the same pleasure in their dispersal.'

Among the many impressive authentic forgeries Bonhams listed in its sale, the masterpieces of two fakers stand out as intriguing genuine fakes – the works of the Spanish Forger and William Henry Ireland. The Spanish Forger painted and sold faux medieval images during the late nineteenth century and William Henry Ireland is famous for his eighteenth-century fraudulent William Shakespeare signatures and invented plays. The celebratas of these forgers and their forgeries explains why they were in the Schimmel collection to begin with, as both have become fashionably and lucratively collectable in their own right. These forgeries prove that fakes can be more than just the stories of them being debunked.

★ ★ ★

'The Spanish Forger was one of the most skilful, and successful, and prolific forgers of all time,' William Voelkle, Curator of Medieval and Renaissance manuscripts at the Morgan Library, contends in his numerous publications about the Spanish Forger's work throughout the late twentieth century. Full of admiration for the Forger's talent and audacity, Voelkle argues that 'Until recently his numerous panels, manuscripts, and single leaves were appreciated and admired as genuine fifteenth- and sixteenth-century works. Now they are increasingly sold, collected, and even exhibited as his forgeries.'

The story of the Spanish Forger begins in Paris. In the latter half of the twentieth century, Paris was not only the centre of influential arts movements, but also a headquarters for forgery production targeting tourist markets in the city as well as agents buying on behalf of high-end collectors and museums. The presence of less than genuine art was ubiquitous, whether that took the form of a copy, a replica or an outright forgery.

The issue of forgeries became so prevalent that in 1904, the French art historian and critic Count Paul Durrieu published a series of scathing articles warning his readers that the markets were rife with fakes. Whatever sort of art people wanted to buy, Durrieu argued, that art was being faked (and some of it was even being faked well). Although Durrieu did not mention any specific works that would later be attributed to the Spanish Forger, it is clear from the pieces that he did highlight that the Spanish Forger would have known how to work the Paris market of fake art. The Spanish Forger recognised just what sorts of fakes would be snapped up by enthusiastic, if somewhat gullible, buyers.

For decades, the Spanish Forger's work slid under the art world's radar, working their way into a plethora of collections, both private and institutional. The odd, newly discovered work still shows up, even today; as recently as 2016, a 'new' Spanish Forger piece was discovered and authenticated by the *Antiques Roadshow*. Although rumblings of fakery surrounded a couple of the Forger's manuscript pages early on in his story – in the mid-1910s – it wasn't until 16 years later that the faker and his phony medieval art were formally recognised in the communities of art authentication.

In 1930, Ms Belle da Costa Greene, director of the Morgan Library, was asked to authenticate *The Betrothal of Saint Ursula*, a purported Jorge Inglés fifteenth-century painted panel. Maestro Jorge Inglés was a well-known painter and illuminator in the mid-fifteenth century, primarily painting out of Castille; one of Inglés's best-known panels is the *Altarpiece of the Gozos de Santa María* (known colloquially as the 'Altarpiece of the Angels'), a panel considered by many scholars to show the intersecting Hispano-Flemish painting traditions. Acquiring an Inglés would have been a coup for any museum, which is why the purchasing agent for the Metropolitan Museum of Art's board of trustees, Count Umberto Gnoli, was eager to add *The Betrothal* to the Met's collection. *The Betrothal of Saint Ursula* had already been authenticated by Sir Lionel Henry Cust, whose opinion Gnoli accepted. In order to shore up the board's confidence in the panel's authenticity, Gnoli hoped that Greene would endorse the panel and recommend that the board acquire it. The asking price was £30,000.

Bella da Costa Greene is an enigmatic character in the Spanish Forger story. Twenty-five years before coming across *The Betrothal of Saint Ursula*, she was hired by John Pierpont Morgan as a librarian to manage his then ever-growing collection of art, antiquities, manuscripts and rare books. Junius Spencer Morgan II – Assistant Librarian of rare books at Princeton University and John Pierpont's nephew – introduced Greene to his uncle and arranged for her initial job interview.

Belle da Costa Greene was born Belle Marion Greener, the daughter of Genevieve Ida Fleet and Richard Theodore Greener, who was (in 1870) the first black graduate of Harvard University. Greene grew up in Washington, DC, in what historian Heidi Ardizzone describes as a struggling elite community of people of colour – well educated, well connected, but still very much situated in the racism of early twentieth-century America. Around 1900, after her father had separated from her mother and taken a post as a US diplomat in Siberia (where he subsequently had a second family), census records show that Belle's family began to vary their names. Before she arrived at Princeton, Belle dropped her middle name, 'Marion', and added da Costa.

When Greene began work at 'Mr Morgan's Library' in 1905, she brought with her the expertise in illuminated manuscripts that she had acquired at Princeton. She started as Morgan's personal librarian at a salary of $75 a month, and over the next four decades would handle millions of dollars of business for J. Pierpont Morgan and, after his death in 1913, for his son Jack. She eventually became the Director of the Library when it was made a public institution in 1924, and steered the collection and the library to be one of the foremost of the twentieth century.

Perhaps most significant to Belle's own biography, however, is her decision to create her own origin myth – her own story of where she came from – and use it to leverage her considerable intellect to her advantage. In her initial interview with Morgan, she said that her mother was of southern nobility, fallen on hard times, and that her grandmother was of Portuguese descent, explaining the da Costa part of her name as well as what newspapers of the time called her 'exotic' appearance. According to Ardizzone, just how Belle's biracial heritage impacted her identity is an open question – Belle famously burned her personal correspondence before her death and carefully controlled her public persona over her 43-year career.

Although Belle was openly extolled as a leading scholar of illuminated manuscripts, she never worked in academia or held a formal university profession, and never published scholarly research or books. She did, however, as one her admirers noted, 'transform a rich man's casually built collection into one which ranks with the greatest in the world'. Greene easily held her own among the intelligentsia and aristocracy in New York City and Europe, and in 1939 she was the second woman elected as a Fellow to the Medieval Academy of America. Greene embraced her reputation as a bohemian librarian, known for her acerbic wit. In one of her most-quoted quips, she resoundingly rejects the dowdy stereotyping of her profession. 'Just because I *am* a librarian,' Belle declared, 'doesn't mean I have to dress like one.'

★ ★ ★

When Count Gnoli sent a picture of *The Betrothal of Saint Ursula* to Belle da Costa Greene in 1930, he had no idea that the panel would kick off of a decades-long hunt for one of the most elusive forgers in art history.

The Betrothal of Saint Ursula is a downright idyllic painting. It's also considerably hefty. After spending months looking at prints and scans of *The Betrothal* in art books or posted online, I had internalised the painting to simply be page sized, despite dutifully typing out its dimensions multiple times. When I put in a request to see *The Betrothal* at the Morgan Library, the librarians wheeled it out on a dolly, carefully propped up on foam blocks. The size of the panel caught me by surprise. It measures 60cm (24in) across and just about 76cm (30in) tall, not counting the massive frame and box the piece is stored in. The librarians told me that they were excited to see it in person, as it's a rarely requested item from the Morgan collection.

In *The Betrothal of Saint Ursula*, lollipop-like trees dot the background and small ships bob around in a harbour.

A Lancelot-and-Guinevere styled couple wend their way from one castle to another. All of the figures have little bow mouths, pursed into syrupy smiles. Ursula demurely offers her hand to her green-sleeved fiancé, and her noble ladies-in-waiting mince down the lane in their bejewelled headdresses and low-cut dresses. White, puffy clouds fill the blue skies and you can't help but get the sense that it's always sunny in Saint Ursula's world. Belle da Costa Greene looked at the painting's medieval halcyonic utopia and concluded that there was something decidedly hinky about the whole thing.

There was something – or some things – that simply didn't sit well with her. It was cracked – with age? maybe? maybe not? – but the cracks conveniently stopped before they destroyed any of the figures in the painting. It was just too … something. Too cutesy. Too perfect. Too many unflawed details that didn't quite add up. Ironically, the panel just looked too medieval.

After a thorough analysis, Greene concluded that *The Betrothal of Saint Ursula* wasn't a Jorge Inglés and that it certainly didn't date to any semblance of the Middle Ages. She jotted down her conclusions on the back of a colour illustration of *The Betrothal* that had been reproduced in *Art News* (1929), where Sir Lionel Henry Cust had originally attributed it to Inglés. 'Brought to me by Count Gnoli, who was then purchasing agent in Italy, for the Metropolitan Museum of Art,' Belle da Costa Greene recorded, 'who asked me to "back him up" in its importance for the Museum when it was to be passed on by the Trustees.' She christened *The Betrothal*'s artist the 'Spanish Forger' – because *The Betrothal* was purported to have been the work of a Spanish artist – put the panel as having been painted at the turn of the twentieth century, and over the next nine years, began to create an oeuvre that carefully documented this faker's faux medieval works. Incidentally, but unsurprisingly, the

Metropolitan Museum of Art declined to purchase the
painting.

Although Greene began the systematic work of
documenting known instances of the Spanish Forger's
work, two of his specific pieces had already raised questions
as early as 1914. Salomon Reinach, the famous French
archaeologist and antiquities expert, exposed one painting,
The Portrait of a Young Lady of Sixteen, as a forgery; later that
same year, Reinach and Henri Omont found another fake
medieval work in a copy of an illustrated Juvenal manuscript.
In both cases, however, the work was simply categorised as
'fraudulent' and remained unattributed to a specific artist.
That changed with Belle da Costa Greene's frank assessment
of *The Betrothal of Saint Ursula*. Now, *The Portrait of a Young
Lady of Sixteen* as well as the Juvenal were attributed to the
Spanish Forger.

It turned out that Greene had seen the Forger's work
before. The Morgan Library itself had one of the Spanish
Forger's works that Greene had authenticated in 1909 –
a medieval liturgy book – and the Forger had even
finagled his way into the collections of the Metropolitan
Museum of Art. Greene's list – which eventually totalled
14 items during her curatorial tenure – has continued to
grow in the ensuing decades, as art experts ferreted out
more examples of the Forger's work among public and
private collections.

'As I have been interested in my artist friend, whom for
want of a genuine name I call "the Spanish forger," I am
delighted to learn from your letter of September 26 that
you have uncovered several others,' Belle da Costa Greene
wrote to Charles Cunningham, curator of paintings at the
Museum of Fine Arts in Boston. Cunningham in 1939, the
like most of the curatorial art world, had learned that
Greene was on the hunt for the Forger and was happy to
help. 'I think that he (or the factory) started in the

manuscript field, although I have seen several paintings undoubtedly by the same hand … I have no doubt that I shall be able to run them all down to manuscript originals. The last one I discovered was really fantastic.'

Curators sent Greene photographs of potential Spanish Forgers, both those in their own collections and ones that were up for auction. 'Thank you for the photographs which arrived this morning,' Greene wrote to Cunningham in a subsequent letter, in late September 1939. 'Yes – they are all examples of my dear old pal.' Two years later, however, the charm of this errant scamp's forgeries was wearing a bit thin. 'I am infinately [sic] ashamed not to have thanked you before this for your letter of September 17th and the accompanying photographs,' Belle da Costa Greene groused to Charles Cunningham. 'The "guy" is becoming rather tiresome, do you not think? Two of these photographs are practically duplicates of two others I have.'

For decades, many have tried to put a name, or at least a definitive nationality, to the Forger. Scholars can trace the Spanish Forger's inspiration to a five-volume series published by Paul Lacroix in France in 1869–1882, which offered detailed illustrations of medieval and Renaissance life; they have even pinpointed which specific Lacroix motifs the Forger used in his various pieces. Using the publication dates of this series to culturally triangulate when the Spanish Forger was actively painting, scholars surmise that his earliest work was in the early 1900s, and decades later scholars would trace the provenances of various Spanish Forger pieces to Paris.

Today, the true identity of the artist remains unsolved – we don't know his nationality, his life story or even if the artist was a man. And that mystery is undoubtedly a key part of the Spanish Forger's continued fascination and appeal.

★ ★ ★

So just what makes a genuine work of art by the Spanish
Forger? Several things, really. Some indications are obvious
to even the most amateur of art enthusiasts (once they
know what to look for), while others are much harder to
uncover, relying on forensic tests and methods.

To begin with, there are the out-and-out saccharine
scenes like those in *The Betrothal of Saint Ursula* that
originally clued Greene in to the fakery. All the faces in the
Spanish Forger's illustrations have a tilt, the mouths are
rather bow-like and all of his subjects' feet turn out as if
they are performing in a ballet corps. Moreover, the Forger
opts to show a very specific sort of secular medieval life,
full of chess and falconry, knights and women in pointy
wimples – these are scenes with gardens, music, games,
unicorns and revelry. Very few of the Spanish Forger's
paintings focus on the religious aspect of medieval life as he
gauged, quite accurately, that his turn-of-the-twentieth-
century bourgeois collectors' interests would be piqued by
scenes that featured castles, jousts, music and dragons, and
not so much the religious side of the Middle Ages.

Another classic giveaway is the Spanish Forger's buxom
noblewomen – ladies with décolletage so daring that they
scarcely manage to avoid serious wardrobe malfunctions.
('All of the women have boobs,' a colleague deadpanned
with a sigh and an eye roll, showing me some Spanish Forger
manuscript pages. 'Any real medieval painting doesn't show
cleavage like the Spanish Forger does.') If one puts all of
these indications together, it really is as if the Spanish Forger
was illustrating the nineteenth century's version of the
Middle Ages. These could be scenes straight out of Sir Walter
Scott's *Ivanhoe*, with all the trappings of medieval chivalry as
could only be imagined hundreds of years later.

The Spanish Forger was careful to use materials that
wouldn't draw a lot of scrutiny from collectors. He stuck
to painting wooden panels, and illuminating loose leaves

and isolated manuscript pages. For pieces like *The Betrothal of Saint Ursula*, the Forger simply used old wooden panels painted on both sides – ostensibly to prevent them from warping. (Some of his contemporary forgers weren't as discerning in their materials and used any convenient source of wood for their medieval-esque paintings. One fake panel from the same turn-of-the-century Parisian market of fakes was discovered to have been fashioned from a desk drawer – the giveaway was a keyhole visible on the back of the panel.)

On close examination of *The Betrothal*, scholars have found that many of those convenient cracks were most likely actually created by the Spanish Forger using a sharp metal instrument. (Rather ironically, the cracks were almost repaired by an art-restoration department before the panel's unveiling as a fake.) Art scholars even traced the source material for *The Betrothal of Saint Ursula* to Lacroix's *Vie Militaire* and Louandre's 'Dames, Nobles et Suivantes', but emphasised that the images are composites inspired by classic texts published at the end of the nineteenth century.

However, it's important to look at *The Betrothal* as a whole object to understand its appeal and success. In the early twentieth century, without forensic testing, it would have been impossible to draw any conclusions from painted wood about the relative age of the material. It *looked* old, ergo it *was* old.

In addition to painting panels, the Spanish Forger forged illuminated loose leaves and manuscript pages by adding his own illuminations to actually genuine medieval manuscripts, thus making his fakes more believable, because they were painted on to something genuine. In a modern examination of the illumination of his painting the *Exorcism of Mary Magdalene*, one of the Forger's few religiously themed pieces, the border of the manuscript page is several centuries old, but the ermine-clad Mary Magdalene is decidedly not. The Forger also scraped many medieval-age vellum pieces and simply added illustrations to them, in essence making these pieces into modern palimpsests where the drawings were often downright incongruous with their accompanying texts.

One such Spanish Forger manuscript leaf, held at the Harry Ransom Center at the University of Texas at Austin, shows Saint Elizabeth greeting the expectant Virgin Mary with three onlookers. The other side of the page shows six sets of four-line red staves with square musical notations. As I sat in the reading room of the Harry Ransom Center and looked carefully at 'The Visitation', I could see the blade marks and bites along the edges of the page. The cuts were just slightly uneven, and it quickly became clear how the Spanish Forger had simply cropped part of an antiphonal leaf from an actual fourteenth- or fifteenth-century musical prayer book, and gone on to paint his nineteenth-century scene on the back.

The Spanish Forger had an idea about what fakes he could get away with, and an unwillingness to tempt himself to create something on a much larger scale. Entire books, for example, would be very difficult to fake well, since the possibility for error in the forgery compounds with the complexity of the object. 'The faking of whole manuscripts [is] attempted but rarely,' Otto Kurz explains in his extensive mid-twentieth-century handbook of European forgeries, geared to students and collectors. 'For every manuscript, however genuine, may contain spurious parts, ranging from whole pages to inserted initials. It is difficult to manufacture whole manuscripts, but it is comparatively easy to effect minor changes, which, if taken at face value, may increase not inconsiderably the interest of a manuscript.' Keeping it relatively simple meant that there was less possibility of the Forger giving away his game. The real genius in the style of the Spanish Forger is his savviness in simply sticking to his own developed genre of medieval fakery.

This doesn't mean that the Spanish Forger never erred on details. When he attempted more complicated multi-panel projects – like the Cincinnati Triptych or the Tewksbury Polytych – it is obvious that both pieces were never meant to be viewed with the side panels closed. In an authentic triptych – a three-part panel painting with hinges on the two outer panels to enable them to open and close – the artwork can be viewed with the panels either open or closed. In the Spanish Forger's triptych project, if you were to close the outside doors on either panel, you would find that the painting doesn't, technically, wrap all the way around – an immediate indication that the triptych isn't genuine.

Over the course of the early twentieth century, art specialists built a profile of the Spanish Forger's style and methods. By the 1970s, authenticators began adding forensic techniques to their forgery-detecting arsenal as new scientific methodologies became available. Could we discern genuine Spanish Forger paintings scientifically, experts wondered?

The illuminated manuscript page *Hunt of the Unicorn Annunciation* offered just such an opportunity to blend art and science – what specialists saw as an opportunity to merge their areas of expertise, offering more insight into the Spanish Forger's art overall. The *Hunt* shows the archangel Gabriel with two leashed hounds facing a kneeling Virgin Mary, who is in turn holding a small unicorn. The entire scene is set amid a lush field of green grasses, a canopy of trees and a golden sky. Art experts argued that the piece was a ridiculously simplified take on a complex religious allegory – the Forger had simply opted to take the most 'interesting' elements of the illustration from the original medieval texts. Tellingly, art experts pointed

out, the Council of Trent strictly forbade the depiction of
the *Hunt of the Unicorn Annunciation* in 1563; a sixteenth-
century artist ought not to have been painting that piece,
in that style, at that time. Again, to the discerning expert
there is a sense that 'something's off'. But the *Hunt* begged
the question, what could science add to the Spanish Forger
story that would corroborate what art experts already
knew?

This is where the forensic testing came in. Art experts
subjected the *Hunt of the Unicorn Annunciation* to neutron
irradiation at Brookhaven National Laboratory in conjunction with the State University of New York at Stony Brook,
under the supervision of the Morgan Library and Museum,
to better understand the choices in paint used by the
Spanish Forger. The results were extraordinary.

The *Hunt of the Unicorn Annunciation* was bombarded
with a beam of slow neutrons, exposing the manuscript leaf
to low levels of radiation. (This sort of test doesn't harm the
original work of art.) The neutrons from the beam stuck to
the different atomic nuclei in the paints, causing some of
the atoms to become radioactive isotopes with different
and distinctive half-lives, depending on the paint's
elemental composition. (Each paint colour contains
different elements and these elements are necessary to give
the colour its specific visual hue.) Scientists took several
X-ray autoradiograph images over the course of 19 days to
see how the different radioactive isotopes in the paint
behaved, and whether there was a paint-based pattern to
how the isotopes decayed. In the case of the *Hunt*, it was
quickly obvious that the green paint in the illumination
decayed faster than the gold, which decayed faster than the
red – the images from the X-ray films left ghostly outlines
of the gold paint, whereas the green paint had faded.

Physicists at Brookhaven took measurements of the
radiation energies from the manuscript page and found

high concentrations of gold, mercury and arsenic. The green pigment, as it turns out, was made from a copper arsenate and was part of a common paint called Paris Green; this was a significant finding because Paris Green was not available before about 1814. Science concurred with curators' expertise that there was no way the *Hunt of the Unicorn Annunciation* could possibly be medieval, a conclusion that was in complete accord with the deductions based on the illumination's subject matter.

Paris Green pigment aside, once you know what to look for it's impossible – even for a non-art expert like myself – to fail to see the various signs that point to the Spanish Forger's work. Looking at *The Betrothal of Saint Ursula* at the Morgan Library, I found myself going through Belle da Costa Greene's original checklist of tells. Little bow-mouths, ballet-like feet, cutesy castles and daring décolletage. Check, check, check and check. Other illuminated vellum pages from the Spanish Forger that I saw offered beautiful miniature scenes, full of knights, battles and the nobility with pointy princess hats. (Reader, I could not help smirking at a noble lady's lips puckered into a trout pout.) But the detail! I was mesmerised by the details in the illustrated miniatures.

I had been prepared to arrogantly dismiss the Forger's work, well armed with all of the Spanish Forger tells, but the skilled detail in each and every page reminded me that the Spanish Forger – whoever the artist was – was a brilliant, regardless of the context of the paintings. When I was sat at a table in the Harry Ransom Center in Austin, Texas, and looked at how the Spanish Forger repurposed and repainted the page of medieval music into a faux-medieval miniature, I could see the hundreds of decisions that resulted in the page being what it was and how it was in front of me. These were decisions made about the original manufacture of the antiphonal prayer book in the fourteenth–fifteenth centuries, and the choices that the Spanish Forger made in creating what I was actually looking at. I couldn't help but wonder if that green of Saint Elizabeth's robe in the painting would light up as Paris Green under neutron activation analysis.

Today, examples of the Spanish Forger's work can be found in many prominent art and manuscript collections, both public and private – Dartmouth, Harvard, the Metropolitan Museum of Art, the Walters Museum, the University of Pennsylvania and the University of Texas at Austin's Harry Ransom Center, to name a few. Even better, more pieces are likely to be discovered. Most estimates put the Spanish Forger's known catalogue raisonné at around 350 pieces and growing, as we don't know how many 'original' Spanish Forgers there are out there – and it leaves us with the possibility of discovering new ones. If the history of the Forger is any indication, art experts are certain to unearth and 'authenticate' more of his work in the decades to come.

It goes without saying that due to their fabulously curious, historically quirky provenience, the Spanish Forger's fakes sell, and sell well, with many pieces fetching astronomical prices at auction, and reserve prices often running in the tens of thousands of pounds. When four Spanish Forger pieces from the Stuart B. Schimmel Forgery Collection were auctioned in 2012, their reserve prices ranged from £4,000 to £8,000. It is apparent that what collectors want is the real forgery and the story of the Spanish Forger to go with it.

★ ★ ★

The story of William Henry Ireland stretches even further back in history, predating the work of the Spanish Forger by more than a century. The immediate nineteenth-century success of the Spanish Forger pivoted on establishing a unique style of giving audiences exactly what they wanted in the media that they desired. Similarly, Ireland's success hinged on his audience's obsession with all things William Shakespeare. The story of Ireland's forgeries speaks to the power and cachet that objects take on – even after they've been thoroughly debunked. Ireland's fraudulent Shakespeares are simply the story of fakes all the way down.

William Henry Ireland was born in 1775 to a rather unremarkable family in London. His father, Samuel Ireland, was an antiques collector, publisher of travelogues and ardent worshiper of the Bard's literary genius. As a rare book collector in his own right, Samuel Ireland desperately wanted to acquire something signed by Shakespeare himself to enhance his gravitas and reputation as a serious collector. In 1794, William Henry was working in an uninspiring job as a legal assistant, and his father dismissed him as too much of a 'dullard' to be interested in the world of collecting. William Henry desperately wanted to prove his father wrong.

Thus, at the age of 19, William Henry forged a simple business deed that bore Shakespeare's signature, using several old rolls of scrap paper that he considered authentic enough to pull off the feat. 'Having cut off a piece of parchment from the end of an old rent-roll at chambers, I placed a deed before me of the period of James the First, and then proceeded to imitate [sic] the style of the penmanship as well as possible,' Ireland wrote several years later. 'This gave me a deed that formed a lease between William Shakpeare [sic] and John Heminge with one Michael Fraser and Elizabeth his wife, whereto I affixed the signature of Shakspeare [sic], keeping the transcript of his original autographs before me.'

When presented with this 'discovered' Shakespearean deed, Samuel Ireland was ecstatic; it never crossed his mind that it wasn't genuine. William Henry quickly forged some more items, which were also successful. And on it went. William Henry claimed to have found the Shakespearean documents through a colleague of his from school, and the story was just credible enough to pass his father's scrutiny as well as that of other, notable experts. Over the next year and a half, from December 1794 until 1 March 1796, Ireland faked William Shakespeare's signature on legal documents, drafted a new 'lost' Shakespearean play (one *Vortigern and Rowena*), 'found' another 'lost' play (*Henry II*) and even crafted entries in 'family Bibles'. Anywhere and everywhere that one might hope to see Shakespeare's name pop up, William Henry Ireland ensured that it did.

Samuel Ireland 'treated his new treasures with appropriate idolatry', as contemporary Ireland expert and bookseller Arthur Freeman describes in his overview of Ireland fakes, written for Harvard University. Samuel exhibited the forged discoveries – the public could buy a ticket sold by Samuel, with scheduled thrice-weekly visiting hours – and offered interested subscribers a selection of the better texts, and facsimiles in the form of a four-guinea folio.

The technique for crafting these forgeries wasn't easy, and William Henry used a mixture of methods to hood-wink his father and the public. As he described in his autobiography, *Confessions*, he would often use bits and scraps of seventeenth-century paper that could pass cursory scrutiny, as he had with that first deed. (This was similar to the method of material sourcing employed by the Spanish Forger.) Make your forgery with enough of the real thing, the logic goes, and it helps suspend disbelief in the fake bits.

William Henry varied the spelling of Shakespeare's name in his faked letters and deeds (Shaxpear, Shakespeare, Shakspere, Shakspeare and so on), which would have been in keeping with the different spellings that Shakespeare himself used – an authentic detail to make the forgeries all the more believable. William Henry also tracked down seventeenth-century quartos, playbills and other pieces that were historically genuine, and then simply add his own signature, signing as William Shakespeare. The method of mixing in real parts to his forgeries was an effective way of keeping up his charade. He also used his forgeries to 'authenticate' previous forgeries, helping him avoid awkward questions about his discovered Shakespearean ephemera and documents.

But procuring the ink for the phony Shakespeares proved somewhat difficult. When he started forging papers, William Henry discovered that Thomas Laurie's bindery – within walking distance of New Inn, where Ireland was cranking out his forgeries – could produce an ink that looked downright Elizabethan by mixing in a little bit of acid to make the ink really froth. When he bought his original vial of old-fashioned ink, he apparently joked with Laurie and his apprentices that he wanted to see if he could use the ink to fool his father into thinking that a document was older than it really was. By the time William Henry needed to refill his inkwell, the fame of his Shakespearean 'discoveries' was well known; yet when he went to the bindery and asked the same assistant to brew him up another batch, no one said anything about anything. 'Although the fame of the manuscripts was perfectly well known to them,' William Henry mused, 'and I was the person supposed to have discovered them.' One can perhaps assume that Laurie and his assistants were so jaded that they simply didn't care what William Henry used the ink for, just as long as he bought it from them. Forsooth, William Henry Ireland paid his shilling for that second vial of ink and kept right on forging.

Samuel Ireland put William Henry's 'discovered' documents together, added his own attempt at a scholarly commentary and published the whole thing as *Miscellaneous*

Papers and Legal Instruments under the Hand and Seal of William Shakespeare on Christmas Eve 1795. Reactions to Samuel Ireland's magnum opus varied, to say the least.

The appeal of Shakespeare cut across all social classes, and offered cultural legitimacy and bourgeois *habitas* to Shakespeare collectors. (One particularly popular item was a love letter Ireland crafted from Shakespeare to his wife, Anne Hathaway, which purported to contain a lock of Shakespeare's own hair.) Everyone loved something about Shakespeare; everyone had something – a character, an experience, something – that they felt could connect them, personally, to the playwright. Ireland's fakes not only fooled members of the English nobility; they even passed the scrutiny of illustrious eighteenth-century scholars like Dr Samuel Parr and Dr Joseph Warton. Upon seeing Ireland's genuine fakes, the learned man of letters James Boswell (friend of renowned playwright Samuel Johnson), was rumoured to have sunk to his knees and kissed the precious relics. The Papers, as they came to be known, were adored by the public, right up until they weren't.

From the start, there were prominent Shakespearean scholars like Edmond Malone, who were dubious of the Papers' authenticity and incredulous that Ireland would have had the good fortune to find so many items that fitted so perfectly with the public's expectations. When Samuel Ireland published *Miscellaneous Papers*, it proved to be one fake thing too many for Malone to handle. In 1796, Malone published a truly blistering book, *An Inquiry into the Authenticity of Certain Miscellaneous Papers and Legal Instruments*, which soundly debunked the Papers as well as William Henry Ireland's stories of discovery. (Malone had scant tact or patience, even, for what he saw as Samuel Ireland's ludicrously amateur publication from the year before.) Malone pointed out inconsistencies in the materials, as well as arguing that their provenances and details were too odd for them to be authentic Shakespearean documents.

For example, there was a set of handwritten marks on Ireland's *Lear* manuscript's final page that were actually doodles, which William Henry had forgotten to clean up. There was also the business of William Henry taking a few liberties with the original text. In the original *King Lear*, Shakespeare had the audacity to have Lear rant the lines: 'I woud divorce me from the mother's toombe/Sepulchring an adultress.' These lines, in Ireland's version, became 'I could divorce thee fromme thye Motherres Wombe/And say the Motherre was an Adultresse ...' Ireland's audiences felt that the second version – the forgery – was a vast improvement, as it helped clarify Shakespeare's pesky subtlety and made Lear's diatribe that much more understandable. *This* is what Shakespeare meant to have written, if only the Bard had had the good fortune to write with the clarity that Ireland imagined Shakespeare could have had. Edmond Malone, however, was having none of it.

Upon Malone's publication of *An Inquiry*, the younger Ireland quickly capitulated and acknowledged that he had set out to dupe his father and the public. In his autobiography, *Confessions*, he wrote that 'it occurred to me, that if some old writing could be produced, and passed for *Shakspear's*, it might occasion a little mirth, and shew how far *credulity* would go in the search for antiquities'.

Incidentally, the Irish playwright Richard Brinsley Sheridan purchased the rights to produce the 'lost' Shakespearean play, *Vortigern and Rowena*, from Samuel Ireland. Sheridan even staged a production of it in the spring of 1796 at the Drury Lane Theatre. The actor-manager John Philip Kemble protested vehemently against putting on the play, as he and other actors simply didn't buy into *Vortigern and Rowena*'s authenticity and never would. It was rumoured in the theatre world that Kemble purposefully repeated the line 'And when this solemn mockery is over' to make sure that audiences knew very well just what, exactly, he thought about the play. *Vortigern and Rowena* opened on 2 April 1796 – three days after Malone published his *Inquiry* – despite Kemble's suggestion that a more fitting opening night would be the day before, April Fools' Day.

'The final two acts went downhill and the audience, stacked with pro-Malone disbelievers, shouted the actors off the stage,' as art historian Noah Charney wryly described the incident in his book, *Art Forgery*. 'It was the one and only performance.' Early on, Sheridan had noted that, compared with the Bard's other plays, *Vortigern and Rowena* seemed awfully simplistic and downright inelegant. But if there was a chance that Shakespeare had written *Vortigern and Rowena*, then, by gosh, the Drury Lane Theatre would perform it.

Despite the overwhelming evidence, however, there were many, including Ireland's own father, who refused to believe that the Papers were fake and that William Henry's confession was real. Most biographers of William Henry Ireland agree that Samuel believed that his son was simply 'too dim and shallow' to create something so accomplished as these forgeries, and that he went to his grave disbelieving his son's confession – choosing to believe in the veritable power of these forged Shakespearean Papers. Other public enthusiasts felt the same way. Shakespeare carried too much cachet, supporters argued, for these treasures to be dismissed as fake simply because William Henry Ireland had confessed to faking them. 'The cult of Shakespeare, of Bardolatry, was well established, but the level of Shakespearian scholarship was not,' author Patricia Pierce Card explained of Ireland's appeal in *The Great Shakespeare Fraud*. 'It had long been suspected [by Ireland's contemporaries] that a cache of Shakespeariana would one day be found, such material having been annoyingly limited until then.

London's literati *wanted* to believe that the "Shakespeare Papers" were genuine.'

Even after William Henry confessed to the forgery, Samuel had the items bound in fashionable Russia leather in three folio volumes – with clasps – and commissioned wooden boxes trimmed in green for three special deeds, including the one that purportedly held the lock of hair attached to the letter to Anne Hathaway. Apparently not satisfied with the public mockery he endured when the forgeries were denounced as such, Samuel went for another round in 1799, when he committed to putting *Vortigern and Rowena* and *Henry II* (another 'lost' play William Henry 'found') to print.

Samuel died, as Arthur Freeman described, 'embattled and bitter', in July 1800, whereupon his heirs put his non-Shakespearean books and relics into auction at Sotheby's in May 1801, all of which sold for a measly £52. The rest of the immediate Ireland family – William Henry's mother and sisters – was anxious to rid itself of this entire affair and sold William Henry's original 170 Shakespearean forgeries to John 'Dog' Dent, MP, a banker, politician and bibliophile, who was rumoured to have paid £300 for them.

But William Henry Ireland's story, like so many stories of fakes, forgeries and frauds, doesn't simply span the 18 months when he was cranking out Shakespeare ephemera. Yes, between 1794 and 1796, he produced those original fake Shakespearean Papers. But it would appear that he wasn't quite ready to let the forgeries go – he was also perpetually broke, and with the ever-present threat of debtors' prison hanging over him, it's not hard to imagine why he opted to pick up his forger's quill again. Incidentally, William Henry found a rather spectacularly brilliant way to capitalise on his celebrity (notoriety?) by *beginning to forge his own forgeries* and selling them as his original genuine fakes.

As early as March 1797, William Henry Ireland sold a set of 10 forged forgeries to his neighbour and lawyer Albany Wallis. For the next four decades, William Henry had a long and, we might say, successful career manufacturing hundreds of artefacts that he often claimed were the famous 1794–1796 'originals'. (Recall that William Henry's mother had sold all of those artefacts to Dent in 1801.) Nonetheless, as rare booksellers note, dozens of these 'copies' exist and surface in auctions – but the provenance and the original forgeries are safely authenticated. However, the forged forgeries sell, and sell well, especially to modern collectors.

'He gave these people only what they were craving, and were eager to swallow,' claimed art historian Richard Grant White, describing Ireland's appeal and why so many bought the hoax for so long. But Grant White's observations of Ireland's appeal are also apropos to why people were as immediately keen to collect Ireland's stuff, just as soon as it was revealed that the Papers were fake. He also describes Ireland's appeal to modern collectors: 'The real inspiration behind Ireland's work lay in understanding the psychology of his audiences, and as they cried for more he fed them according to their appetite and capacity of belief.' The 2012 auction of the Stuart B. Schimmel Forgery Collection boasted several Irelands – everything from letters signed as himself, 'WH Ireland', to what was actually authentic Shakespearean ephemera that happened to have Ireland's signature as 'William Shakespeare'. Incidentally, but unsurprisingly, Ireland signing as Shakespeare outsold anything signed as 'WH Ireland'.

Today, William Henry Ireland is one of the most collectable and valuable of the dubious pantheon of Shakespearean forgers, with authentic Irelands fetching anywhere from a few hundred pounds to tens of thousands of pounds at auction. More than just financial cachet, though, William Henry Ireland's forgeries continue to intrigue scholars. Institutions that have Ireland holdings – like Harvard University – consistently find their Ireland forgeries in use for any number of scholarly research projects.

★ ★ ★

When Belle da Costa Greene determined that *The Betrothal of Saint Ursula* wasn't a panel by Jorge Inglés and that it certainly wasn't medieval, the Metropolitan Museum of Art declined to buy it. The panel passed through the hands of private collectors for decades, and was eventually gifted to the Morgan Library in 1988 by Martin W. Cooper, in memory of his mother. Today, the panel is an important artefact of Spanish Forger history, and it is rather fitting that the Morgan Library has ended up as its eventual home.

One of the key parts of the Spanish Forger's story (as well as that of William Henry Ireland's shady Shakespeares) hinges on the moment when the art stops becoming something that is merely exposed as 'fake', and begins to become identifiable and collectable in its own right. In 1978, the Morgan Library put together an exhibition that focused specifically on the Spanish Forger's art in its own right – when American art critic Hilton Kramer reviewed the exhibition for *The New York Times*, he referred to the Forger as a 'nineteenth-century artist'. The Morgan Library had hoped that the exhibition would encourage collectors to re-evaluate their own collections and possibly bring more of the Forger's works to light.

The Morgan Library was not disappointed. When Lady Jeanne Campbell Cram – Norman Mailer's first wife – saw the exhibition, she took a second look at an elaborately framed painting of a *Noblewoman Embarking a Ship* in her family collection. She immediately realised that she had an authentic Spanish Forger. ('Eureka,' she told William Voelkle over the phone, 'I've got one.') Lady Jeanne's grandfather had bought the painting in Paris in the 1920s, and when she brought the painting in its ivory frame to the Morgan for Voelkle's inspection, it quickly became clear how iconic and distinctive the Forger's style is – a non-expert could discover a piece in her collection based on just viewing one art exhibition.

When the catalogue for that Spanish Forger exhibition was going to print, a Californian dealer brought a book of hours, which had formerly belonged to the actor John Barrymore. (A book of hours is a devotional that contained prayers, psalms, and some scriptural text; books of hours were popular during the Middle Ages and often were lavishly illuminated.) Even though the dealer was disappointed that it wasn't a fifteenth-century artefact – although the paper and borders were genuine – Voelkle was thrilled to find that the famous French knight depicted in the book was painted by none other than the Spanish Forger. 'In 1984 John Barrymore, Jr., came up with the idea of a movie script on the Spanish Forger. Naturally he would play the lead, since his father had owned the manuscript, which is now in the Walter Art Museum,' Voelkle reminisced in his book, *The Spanish Forger*. 'The likes of James Coburn, Peter O'Toole, Richard Harris, and Alec Guinness were to be approached as possible participants. For years I had hoped that the film would become a reality.'

These are the sorts of stories that build up the authenticity of the Spanish Forger's art work and make it more real than any sort of validation within the art world. The object – the panel, the book of hours, whatever – had a complex material history that is all its own. Year after year, painting after painting, manuscript after manuscript, slowly the Spanish Forger had become legitimised, and we're more and more willing to move the Spanish Forger's art a little closer to the 'genuine' or 'real' side of authenticity's continuum. There's little doubt that part of what makes stories like these so compelling is the audacity required to pull off such spectacular fakes, and the ingenuity and expertise essential to unmask them.

Originally, these art forgeries were intended to misguide – they tried to convince potential audiences of their legitimacy. In the world that they live in now, however, they're more than just debunked art pieces – they're material objects with complex histories that have lived longer as 'Spanish Forgers' than they ever did as pieces passing for authentic medieval works. Similarly, history has categorised Ireland's Shakespeares – both the original forgeries and William Henry's own genuine fakes – as forgeries much longer than they were ever considered real. (Ireland's unique twist, was his forging of his own forgeries.)

The Spanish Forger and William Henry Ireland are just two characters in the long history of art forgery, but they neatly bookend how genuine fakes have become collectable in their own right by giving audiences exactly what they say they want. They become really 'real' at the point when they are no longer trying to fool anyone. 'There is an illusionism to art – and to its authenticity,' Noah Charney suggests in *Art Forgery*. 'Sometimes the line between masterpiece and forgery is slender or even invisible. For a crime to be committed, someone or something must be victimized or damaged, whether that is someone specific – a buyer who was duped, for example – or a more abstract harm, such as to an artist's reputation.'

For today's audiences, with little risk of being deceived by either artist, the stories can either celebrate the fakers' ingenuity and artistic prowess, or they can serve as cautionary tales to the over-enthusiastic art collector. But the stories also build the fake's provenance, ensuring that collectors want the fake that is real – genuine – and not a knock-off of a knock-off. Consequently, collectable fakes are about rarity and provenance, both of which depend on the logic of desire and economics, not necessarily reason.

'What would the forger have thought about this ironic state of affairs?' William Voelkle mused. 'Would he have been pleased that his works were now collected as forgeries in their own right, and that they could even be sold for more as his own forgeries than as genuine medieval works of art?'

Thus, works by the likes of the Spanish Forger and William Henry Ireland are thus all the more genuine for having been fake in the first place. They are fakes that have become the Real Thing.

CHAPTER TWO

The Truth About Lying Stones

On 31 May 1725, Dr Johann Bartholomew Adam Beringer, Dean of the Faculty of Medicine at the University of Würzburg in Bavaria, was given three fossils that had been collected from the slopes of the local hill, Mount Eibelstadt. Two of these stones looked very much like a set of smiling worms and the third was a figure of the sun, complete with rays of light shooting out behind it. These fossils, Beringer quickly and confidently concluded, were no ordinary fossils. And, in fact, history would soon reveal that they weren't really fossils at all.

An ardent naturalist and fervent amasser of all sorts of natural curiosities, Dr Beringer employed a host of Würzburg locals as labourers to collect fossils from Eibelstadt. Over the years, he had filled his personal cabinet of curiosities with local fossils, as well as ones even more uncommon than those found in the Bavarian region. 'I acquired the rarer and choicer specimens since I could not find them in Franconia, from almost all regions of Europe,' Dr Beringer boasted in his eighteenth-century writings, 'either by buying or begging, or by the kindness of my friends and supporters.'

The story of Beringer and his not-real fossils began, then, in May 1725, when three of the 'diggers' he employed − 17-year-old Christian Zänger and the teenage brothers Niklaus and Valentin Hehn − brought the three newly discovered curiosities from Mount Eibelstadt to Beringer's door. Beringer was ecstatic. These fossils, he quickly and confidently surmised, looked so very different from the specimens that normally came

from the Eibelstadt region. (He had plenty of seashell
fossils and the like, but nothing as grand as a sun with
sunbeams.) Beringer called the discoveries iconoliths –
lithology was the study of rocks in the eighteenth
century – and encouraged his entourage of employees to
systematically work the scree-filled slopes of the hill in the
hopes of finding more such treasures.

Over the following summer months, Mount Eibelstadt
proved to be a most productive area. Zänger, the Helm
brothers and Beringer's other excavators found fossil upon
fossil, marvel upon marvel. The original three turned into
tens of fossils; tens of fossils turned into hundreds, and
hundreds became thousands. Beringer was practically
beside himself. Historical estimates suggest that by the time
he published his book about the discoveries, *Lithographiae
Wirceburgensis*, in the spring of 1726, he had acquired
somewhere between 1,100 and 2,000 'fossil' specimens.
And all the discoveries were as curiously spectacular as
those first fossils.

In *Lithographiae Wirceburgensis*, Beringer reverently
catalogued his burgeoning collection:

> Here, representing all the kingdoms of Nature, but
> especially those of animals and plants, are small birds
> with wings either spread or folded, butterflies, pears
> and small coins, beetles in flight ... bees and wasps ...
> worms, snakes, leeches from the sea and swamp, lice,
> oysters, marine crabs ... frogs, toads, lizards, cankerworms,
> scorpions, spiders, crickets, ants, locusts, snails, shell-
> bearing fishes ... shellfish, spiral snails, winding shells,
> scallops and heretofore unknown species.

Without pausing for breath, Beringer continued:

> Here were leaves, flowers, plants, and whole herbs ...
> Here were clear depictions of the sun and the moon,

of stars, and of comets with their fiery tails. And lastly, as the supreme prodigy commanding the reverent admiration of myself and of my fellow examiners, were magnificent tablets engraved in Latin, Arabic and Hebrew characters with the ineffable name of Jehovah.

Short of inventorying Noah's Ark, it's hard to imagine a more complete register of life on Earth than what the slopes of Mount Eibelstadt offered to Johann Beringer. Discovering fossils that bore an uncanny resemblance to a sun with a human face, stars, entire frogs that looked rather life-like and, of course, curios that spelled out the name of God quite put something like a fossil seashell to shame. Moreover, with fossils that referenced Jehovah himself, these discoveries were not confined to the mere natural world. Beringer argued that Jehovah's name appeared on rocks as, 'light may somehow have "absorbed" the Hebrew characters from the gravestones (found nearby) and transferred them to the iconoliths of Mount Eibelstadt'.

Beringer was so set on introducing his iconoliths to the world as soon as possible that he took several shortcuts in his analysis. Truth be told, he basically skipped over any sort of study that would have been *de rigueur* in the early eighteenth century. He opted to not consult other naturalists – that is, other experts – about these strange discoveries. Nor did he create a systematic taxonomy of the fossils, or even formal descriptions of what he had discovered.

What Beringer did do, however, was to immediately begin to work on what would become his monograph, *Lithographiae Wirceburgensis*. He even commissioned 21 engraved plates that would illustrate the fossil frogs, shells, birds and comets in his collection. Detailed illustrations of specific specimens under study were an important part of communicating one's observations to one's audiences, and had been since the sixteenth-century Swiss naturalist Conrad

Gesner published *A Book on Fossil Objects, Chiefly Stones and Gems, their Shapes and Appearances* in 1565.

In addition to cataloguing the discoveries, Beringer described several theories about the origin of fossils in general – a question very much of interest to natural historians at the turn of the eighteenth century – thus inserting himself and his iconoliths squarely into active scientific debates. One of the most popular explanations put forward by naturalists was the diluvial theory, which posited that the biblical flood that 'baptised' the Earth drove animal and plant remains into the rocks with the torrential force of its currents and waves; in *Lithographiae*, Beringer spent several pages recounting diluvialism for his readers. Beringer also considered the spermatic theories of Edward Lhwyd and Charles Lang, which suggested that tiny sperms of seeds externally fertilised rocks and proceeded to germinate once inside the stones. He pointed out that discoveries like those from Eibelstadt could go a long way to helping resolve the debate about the origin of fossils.

Despite his unabashed enthusiasm for his trove of fossils and his commitment to his work, Beringer found the book to be hard writing. 'I must confess that I often put down my pen, distracted and worn out as I was by the demands made on me day and night by my duties,' he wrote in the book's conclusion, 'which left me scarcely more than the small hours of the night to compose a work that would be subjected to critical examination by those outside our land.'

Although Beringer never questioned the authenticity of fossil animals that smiled and spelled out the name of Jehovah, other people at the university began to talk. There were whispers that the fossils had a human rather than divine hand in their making. One of Beringer's colleagues at the University of Würzburg, J. Ignatz Roderick, professor of geography, algebra and analysis, even carved a bit of limestone in front of Dr Beringer showing him how easy it could be for someone in the here and now of 1725 to craft the stones, then leave them on the slopes of Mount Eibelstadt for Beringer's workers to find.

Obdurate and tenacious, Beringer remained ever faithful to his fossils. For months, he refused to be swayed by detractors like Roderick and Roderick's colleague, university librarian and privy councillor Johann Georg von Eckhart. Roderick and von Eckhart pointed to the knife marks on the rocks around the carvings, and questioned the probability of discovering something like a spider's web so perfectly preserved. Beringer took their qualms in his stride and suggested that those rumourmongers were merely jealous that they had not had his own good fortune to discover and catalogue such treasures of the natural world.

Boldly, Beringer proclaimed in *Lithographiae Wirceburgensis*, 'I have not given way [to these rumours], as the crowning work of this first edition clearly proves. Further, I shall not yield in the future. I am as determined to champion this most righteous cause, as I have been prepared to throw open this new stone collection of Franconia to the whole world.' What luck – what righteous luck! – on Beringer's part to have been the one to find, describe and publish these fossils.

Except, of course, that they weren't fossils at all and never had been. The iconoliths were fakes and, truth be told, not even particularly compelling fakes at that. Alas, Dr Johann Bartholomew Adam Beringer had been duped by a hoax from the moment he had accepted those first 'fossils' from Christian Zänger in May 1725. The perpetrators of the ruse – the eminent Professor Roderick and von Eckhart, the very colleagues who'd tried to dissuade Beringer from his belief in the fossils' authenticity – had simply thought that they'd have a good laugh at

Beringer's expense and humble him a bit, and that the matter would be quickly resolved.

Little did they know that 12 months, a plethora of fake fossils and two court appearances later, the story of Beringer's *Lügensteine* – his Lying Stones as history would be quick to call them – would prove to be anything but simple.

★ ★ ★

In the first part of the eighteenth century, fossil discoveries were relatively few and far between. While these natural curios had, of course, populated the cabinets of curiosities among the wealthy and educated since the Middle Ages, there were rather few types of specific fossils known among even eighteenth-century collectors. Fossils that showed up in naturalists' collections included the shells of molluscs (like ammonites and belemnites) and sharks' teeth (called glossopterae), as well as a plethora of corals, and plants like crinoids, echinoids and ferns. Perhaps there was the odd vertebrate bone here and there, but these skeletal bits were rare. While many naturalists ascribed these obviously sea-related curiosities to vestiges of the biblical story of Noah's flood, many scholars were interested in building even more robust explanations around fossils and had Beringer's haul been actual fossils, they would have been quite the coup for the burgeoning field of natural history.

It might require quite a bit of suspended disbelief on the part of twenty-first-century audiences to accept that Beringer didn't immediately recognise the artificial and manufactured nature of the stones – artefacts on which knife marks were amply evident, and where the artistry of the stones' carving can be most charitably described as somewhere between 'enthusiastic' and 'rudimentary'. This begs the question of how and why Beringer was taken in by the fossils in the first place. Just what *were* fossils – legitimate fossils – to him and other eighteenth-century naturalists? And how did Beringer's curios differ?

Among Beringer's contemporary naturalists, 'fossils' had a variety of meanings, and many things were described as 'fossils' that simply would not be fossils in contemporary palaeontological parlance. Historically, many naturalists used the term exclusively to describe stones that resembled plants and animals. A few expanded the term to include what would be called archaeological finds today. The word 'fossil' came from the Latin *fossa* and simply referred to objects that had come from the ground. This could include rocks, coins and 'figured stones' that looked like plants and animals, as well as gemstones; this is certainly what comprised 'fossils' collected by seventeenth-century scholars – well before Beringer's time.

'The problem was not to decide *whether* "fossils" were organic in origin or not, but to decide *which* were the remains of organisms (or parts of them),' eminent British geologist and historian of science Martin J. S. Rudwick explains in *Earth's Deep History: How It Was Discovered and Why it Matters*. 'The question was, in which "fossils" was the resemblance to plants or animals due to their origin as part of such living beings, and in which others was any resemblance accidental or a matter of chance?'

This gets at the very question of how history, antiquity and nature act together, and how these three things were foundational in the then-developing discipline of natural history. 'Only those ['fossils'] that were truly organic in origin could be regarded as nature's own antiquities, and therefore be used to supplement, or even replace, other forms of evidence about the *history* of humanity and its terrestrial environment,' Rudwick notes. In other words, by the time Beringer was collecting his genuinely authentic fossil seashells as well as the faux Eibelstadt artefacts, fossils were generally taken to refer to organic remains of things that were not made by humans, although the origin of fossils was still very much under debate. In 1735 – 10 years after the story of Beringer's fossils – Swedish naturalist Carl Linnaeus published the first edition of *Systema Naturae*, offering naturalists a systematic means of biological nomenclature and a method for organising the natural world. Linnaeus's classification system helped to eliminate some of the non-biotic things that Beringer (and other naturalists) classified as 'fossils', since these extraneous objects couldn't be tied to life. Scattered and eclectic though Beringer's catalogue appears, it was nominally in keeping with scientific practices of the 1720s.

Perhaps the best explanation for Beringer's 'fossils' would take them to be artefacts of a much earlier understanding of the natural world – vestiges of the sixteenth-century concept of fossils. For earlier Renaissance philosophers, the universe was a grand, straightforward hierarchy – with God at the top, followed by the angels, followed, in turn, by man, animals, plants and, finally, minerals; in short, the Great Chain of Being in which each thing is a concrete link between something above and below it. With these philosophical underpinnings, the sorts of 'fossils' that Beringer's lackeys found would have neatly and inextricably slotted into such a world view, and the fossil 'links' would have felt familiar and unexpected. The hoax fossils created by Roderick and von Eckhart would have offered Beringer fossils that were exactly what he hoped he would find, whether he consciously acknowledged it or not.

★ ★ ★

Leaving the origin of his 'fossils' as an open question, Beringer devoted himself to publishing the Eibelstadt specimens. At this point, in late 1725, the hoaxers began to worry about the lengths to which Beringer had bought into their scheme. Roderick and von Eckhart had put together the hoax in the first place because they found Dr Beringer to be absolutely insufferable. In their own words, he was 'arrogant' and 'rather despised them all', and they had devised a plan to salt the slopes of Mount Eibelstadt in order to bring him down a peg or two in his own estimation of himself.

Their idea was that Beringer would take a bow about his fossil-collecting prowess, eventually discern that the limestone fossils were fraudulent, then simply feel a bit foolish, once he realised that he had been filling his cabinet of curiosities with fakes – especially if Roderick and von Eckhart had the good fortune to be the ones to point out his folly. (They rather hoped he'd do this in front of someone famous, like their joint patron the Bishop and Duke of Franconia.) However, the preparations for publication of Beringer's book indicated that he had bought into the fossils completely, and now the stakes of the prank were much higher than they had initially anticipated.

When Roderick showed how easy it would be for someone to carve stones like the very ones Beringer found, it was almost as if Roderick hoped that Beringer would pick up on the hoax – as if Roderick was spelling out the ruse for Beringer to put together. Beringer agreed with the idea in principle but refused to consider that his 'fossils' had been so manipulated. Sometime between May 1725 and the actual publication of his fossil treatise in 1726, Roderick and von Eckhart decided that the joke had gone quite far enough and began circulating rumours that the fossils were fake. Beringer inserted a very short, clipped and somewhat snippy chapter towards the end of *Lithographiae Wirceburgensis* that quickly dismissed the idea that these were human-made fossils, and suggested that any claims to such were simply the result of professional jealousy and envy. A few months later, however, Beringer was backpedalling.

The exact date and circumstances of the hoax's unveiling are a little unclear. Some suggest that His Grace, the Bishop and Duke of Franconia, himself stepped into the academic melee and sorted things out, telling Beringer that the rocks weren't fossils. Others claim that Beringer realised that he'd 'been had' when he came across his own name carved in the rocks while surveying the slopes of Eibelstadt. (This is certainly the story of Beringer's Lying Stones that has been mythologised in the history of science.) Although it's impossible to know for certain how the hoax was uncovered, there is little doubt that its effects were immediate.

First and foremost, Beringer was so mortified by his error that he immediately began buying all copies of *Lithographiae Wirceburgensis* that he could get his hands on, in order to head off the possibility that someone, somewhere might actually buy the book, look at the engravings and descriptions, and conclude that Beringer was a total ignoramus for ever having thought that the carvings were real fossils. Then Beringer proceeded to take legal action against Roderick and von Eckhart.

Ostensibly, Beringer felt that the deplorable actions of both Professor J. Ignatz Roderick and Georg von Eckhart had besmirched Beringer's good name and academic integrity. He was determined that history would not remember him as a fool for being duped by the fossils. On 13 April 1726, judicial proceedings began at the Würzburg Cathedral Chapter – at the special request of Beringer – with the hopes of 'saving of his honour'. The judicial process took place over three days, with the last two hearings being held at municipal trials on 13 April and 11 June 1726.

On 13 April 1726, the court heard that Dr Beringer charged 'certain young people of Eibelstadt' who had purposefully brought carved, fraudulent stones to him – as well as 'sundry others' within the natural history collecting community – and passed them off as the real thing. Since Beringer, in turn, had argued for the stones' authenticity, he felt that the question of his own honour was wrapped up part and parcel with the legitimacy of the Eibelstadt rocks. Documents associated with the proceedings referred to Dr Beringer as the Duke's personal physician as well as a 'learned dilettante', lending a smidgen of sympathy to the question of why Roderick and von Eckhart would have found Beringer so insufferable. The court felt that Beringer made a reasonable argument and undertook his case, promising to examine those involved in the hoax as quickly as possible.

Transcripts of the court proceedings suggest that the court's primary aim was to pin down just who had masterminded the entire affair and to what extent the ancillary characters caught up in the hoax knew what role they played in perpetrating the ruse. It quickly became clear, for example, that the 'young diggers' were simply pawns. They excavated the slopes, rounded up the fantastic 'fossils', and were eager to make a batzen or two for their efforts (court records show that the top price for finding one of the *Lügensteine* in 1725 was 22 batzen – roughly the then-equivalent of two Swiss francs). These workers simply collected what they were told to collect by the then 17-year-old Zänger.

Interrogation of the Young Diggers

[COURT]: Did any of them ever learn the art of sculpting?

[COURT]: Did they ever see in any book representations of the figures and characters found in these stones?

[COURT]: Did they ever see anyone secrete such
          sculptured stones on the mountain, and did
          they not dig up such stones?

[COURT]: Were they ever induced by Messrs. Eckhart
          and Roderick to sculpt stones, then to bring
          such stones into the city and pass them off as
          discovered and dug up?

The answer to all of these questions was no, and the court
concluded that no fault or mischievous, malicious intent
lay with the diggers. The questions that the court put to
Christian Zänger were even more direct.

Interrogation of Christian Zänger

[COURT]: Where did Zänger get the 'dragon-stones',
          the stones on which were the Bebraic letters
          and others upon which the knife was used?

[COURT]: Was he not ordered by Roderick or Eckhart to
          deliver up diligently such stones to Beringer?
          To whom else were they delivered?

[COURT]: Whether Zänger does not declare that he
          spent entire weeks at Eckart's house grinding
          and making stones?

[COURT]: Did not Zänger receive from Roderick and
          Eckhart a little sketch of a mouse and Hebraic
          letters? For what purposes were these given
          him?

[COURT]: Did he [Zänger] ever work in alabaster?
          What was the figure he made of it, which he
          described to Eckhart and Roderick?

Zänger wasted no effort in saying that he took his orders
from Herr Roderick and ferried the carved goods from
Roderick's workshop to the hill slopes, and that the other
diggers were simply finding them in good faith. Despite

Roderick's feeble efforts to pin the scandal on Niklaus and Valentin Hehn, the brothers came across as rather haplessly earnest.

[COURT]:   Did [Niklaus Hehn] polish or carve a little, or do anything else with an instrument, or any of the stones he discovered and subsequently delivered to Dr Beringer?

[ANSWER]: He [Niklaus Hehn] never did anything to the stones, but such stones he found he delivered to Dr Beringer *as* he found them – just as he had long before brought him shell-stones.

[COURT]:   Did ... Herr [Roderick] say anything further or make any other requests?

[ANSWER]: He [Roderick] said that his Grace had sent him out to discover whether they themselves [the Hehn brothers] had not carved the stones and hidden them in the mountain. To this he [Hehn] answered that they had done no such thing, for they gave out the stones as they had found them. As Roderick now saw that he could do nothing with him [Niklaus Hehn], he turned to his younger brother to ask him the same question. He also threatened to have him put into irons and getters if his brother would not admit to having carved the stones with a knife.

And so forth. The court even called an innkeeper's wife to testify about Roderick's trips to the Eibelstadt area. More questions were put to Christian Zänger on 11 June 1726, when the court threatened repercussions if it felt he wasn't being entirely truthful. It quickly became apparent that Roderick and von Eckhart had masterminded the entire escapade, and that Zänger knew very well that Roderick was carving the 'fossils' in his workshop because he, himself, was leaving the stones on Eibelstadt for the Hehn brothers and others to discover. Everyone else, to all intents and purposes, it would seem, was just along for the ride.

Then, as abruptly as the hearings had begun, they ended. Four years later, by 1730, the youths and diggers disappear from the historical records, and Roderick had left Würzburg. It's a bit unclear as to whether he left of his own accord, or was more or less run out of town; he did slink back to Würzburg to be allowed access to the city's archive in 1735 to finish a book he was writing when the trial started. Von Eckhart died. Beringer himself continued in his university studies and medical practice, published two additional books of some academic acclaim, and spent the rest of his life hunting down errant copies of *Lithographiae Wirceburgensis* to avoid the book accidently falling into readers' hands.

★ ★ ★

Collecting and buying fossils is nothing new. In the centuries before and after the episode of Beringer's Lying Stones, fossils have been imported and traded as Europe's wealthy have curated their cabinets of curiosities. Corals, fossils and narwhal tusks were extremely popular pieces in these earliest collections, and spectacular fossils have always found a market for the discerning collector. Certainly, fossils have been bought, sold and collected in the United States for more than 200 years; at his Monticello residence, Thomas Jefferson famously kept a personal collection that included a mastodon jaw and teeth.

Part of the mystery of these natural objects comes from their inherently fragmentary nature, and that mystique certainly surrounds fossils today. A fossil forms when an organism dies and its body mineralises over millions of years. It's rare for an entire organism to be perfectly preserved – bones can get trampled on and buried over time, and for a fossil to be captured exactly in situ, the perfect circumstances of geology, geomorphology and taphonomy need to be available. At this point, a fossil is a natural object. Whether an impression of a leaf left in eons-old mud or a dinosaur bone from the Jurassic, it's simply the result of geological processes unfolding through time.

Because of the fragmentary nature of the fossil record, the 'whole story' of an organism is not preserved, just as the 'whole organism' is rarely found. Finding a fossil means recovering the partial remains of a once-living creature from its geologic resting place; giving a fossil meaning requires that science, natural history, folklore and art come together to offer the pieces a coherent narrative. This makes fossils the Real Thing. 'What are fossils, after all,' mused French palaeontologist Pascal Tassy in an interview with historian of science Adrienne Mayor in her book, *The First Fossil Hunters*, 'if not vestiges both preserved and destroyed by time?'

The history of fossil excavation is chock-full of different ways in which researchers and amateurs alike have made these attempts to fill in the gaps in the fossil record. Sometimes this filling in is figurative, but sometimes it's literal. Sometimes a scientific hypothesis is added to round out an evolutionary narrative. At other times the gaps are reduced when a museum fills in a skeleton with copies or casts of bones that weren't recovered with the original specimen, in order to offer visitors a more complete picture of an organism. There are even times when a fossil is so full of gaps, physically and narratively, that scientists and private collectors see what they want to see. And the less scrupulous are more than happy to oblige the eager with genuinely fake fossils.

Once a fossil has been discovered it begins to take on a life of its own. A fossil's cultural cachet comes after it's been found and been given an audience, or even several audiences – be they scientific or commercial ones. It becomes scientifically significant because science is a cultural activity that happens around the fossil. At least, this is the notion of where value comes in, where value is assigned to a fossil in terms of legal and financial circles, leaving the question of a fossil's intrinsic value – be it as a luxury good or a relic of natural history – wide open.

Just as with art and antiquities, almost as soon as there was scientific and commercial interest in fossils, fakes begin to creep into collections. 'Forgers are enterprising,' historian and art crime historian Erin Thompson wryly notes in her book, *Possession: The Curious History of Collectors from Antiquity to the Present*. 'Why not be, in a field where the rewards are high and the punishments are low and unlikely to be brought to bear by buyers who are more willing to cover up embarrassing mistakes than make known the presence of forgeries in their collections?'

★ ★ ★

As early as the Middle Ages, fossil collectors worried about the authenticity of their cabinets of curiosities. For centuries, fossils have been faked in a number of ways and for a variety of reasons. The question of fake fossils – and how to ferret them out from real ones – is still very much a problem for contemporary palaeontologists, museum managers and collectors. While the Beringer case involves one of the most spectacular examples of phony fossils, there are plenty of other fakes that pop up in subsequent centuries.

Fake fossils can be created in any number of ways. Although Georg von Eckhart and J. Ignatz Roderick carved Beringer's Lying Stones out of non-fossiliferous rock, the most historically pervasive sorts of fake fossils, however, cobble together bits and pieces of authentic fossils to create an organism that never lived in nature. These Franken-fossils, if you will, are targeted at providing an audience with exactly the fossil it wants to see. From fraudulent fossils put out into the scientific community as embarrassing hoaxes, to mashed up fossils introduced to collectors' markets to increase the likelihood of profit, most fake fossils start with some small bit of something real.

To begin with, amber-encrusted fossils have been an ever-enduring fixture in scientific and commercial circles for several centuries. 'Collectors (including professional and amateur palaeontologists) compete for acquisition of the rarest, most exceptional amber fossils. Convincing forgeries are relatively easy to make and the financial and scientific stakes can be quite high,' David Grimaldi, Alexander Shedrinsky, Andrew Ross and Norbert Baer argued in an academic journal article in 1994. 'Forgeries were presented as gifts to medieval royalty, and the problem persists today.' Although high prices make these *Jurassic Park*-like fossils some of the most lucrative to fake, they are also among the most easily scrutinised and identified.

As early as 1891, German botanist and naturalist Georg Klebs felt that there was a problem with a skeletonised lizard in the amber collection held by the Becker Museum in Germany. (The fossils were later accessioned to the Königsberg Geological Institute.) If the amber that the fossil lizard was encased in was authentic Baltic amber, it would be the oldest known example of its particular lizard taxa. Klebs conferred with his colleagues and, in 1910, published his concerns and raised the question as to whether the amber that encased the lizard was what it purported to be. In an effort to determine whether the specimen was genuine, Klebs broke open the amber and examined the lizard. (It was then exhibited for the next 25 years in the Becker Museum, including even underwater, for reasons that are lost to history.) Tragically, the specimen was reported missing in the Second World War, and the question of the lizard in amber's authenticity is an open one. This particular piece of amber prompted other museums across the world to reassess their collections of fossils in amber.

Amber fossils can be faked in two ways. Genuine amber is so hard that it is impossible to 're-melt' it in order to encase any sort of plant or animal curiosity in it. Other types of almost amber-like resin (namely copal) are pliable and can be used to create amber-like specimens and visually, copal is practically indistinguishable from amber. It's worth emphasising that copal resin does catch biota in its stickiness, and it's common for pieces of copal to have insects, plants or animals trapped in them; the copal pieces are just not as old as their amber counterparts. These sorts of fossil fakes are cases of simple substitution – giving unsuspecting buyers something that is cheaper and easier to come across than genuine fossils in real amber.

The other method of amber fakery is much more crafty. To make these fake fossils, authentic pieces of amber are opened up, then hollowed out, and a piece of something (plant, animal, insect, whatever) is placed in the amber chamber before everything is glued back together. As improbable as it sounds, this method of creating fake fossils is spectacularly successful. In fact, one such amber fake containing a small, perfectly preserved fly duped curators and staff at the Natural History Museum in London for more than 70 years. This piece of amber had been cut in half and the corpse of a latrine fly (*Fannia scalaris*) had been put into the little carved-out indent. The piece was neatly put back together, and for decades led scientists to think that a nineteenth-century fly had origins that stretched back over hundreds of millions of years. The fake was discovered in 1993, when then-graduate student Andrew Ross was looking at the specimen under a hot light and, to his utter horror, realised that it was melting – the glue that held the two amber pieces together was oozing a bit.

However, this type of fake isn't only produced with fossil amber. One of the most celebrated and notorious fake fossils comes from the field of palaeoanthropology – the study of human evolution.

In February 1912, antiquarian collector and country lawyer Charles Dawson uncovered fragments of a most curious fossil human ancestor in the roadside gravels of Piltdown in East Sussex, England. The fossil pieces looked like those of a hominin – a fossil human ancestor. Upon scientific analysis in the early decades of the twentieth century, the bits of skull and jaw showed a bizarre mix of anatomical characteristics – some human, some decidedly ape-like – leading dozens of researchers to conclude that the Piltdown 'fossil' was the perfect missing link between ape and human in *Homo sapiens*' family tree. The find also championed a model of evolution, researchers argued, that was very much en vogue in the early twentieth century – the hypothesis that humans' 'big brains' were the underlying driving factor in hominin evolution. (And with big brains, the scientific argument went, they were biologically endowed to create a uniquely human culture.) In the 1920s, the Piltdown discovery was touted as the earliest Englishman, and it was held up for decades as incontrovertible proof of the antiquity of ancient humanity in England, although the 'fossil' remained highly controversial outside England.

Piltdown maintained his ancestral status for just over four decades. By 1953, scientists at the Natural History Museum in London had exposed the 'fossil' as a hoax – a fake of the first degree. As a result of batteries of tests, using specific chemical analyses that weren't available to science when the specimen was first discovered, researchers determined that not only was Piltdown not a missing link, it wasn't even a real fossil. The Piltdown skull had been cobbled together from human cranial fragments dated to the Middle Ages, and paired with the jaw of a modern orang-utan, with chimp teeth stuck into the tooth sockets. Certainly, then, it's little wonder that early twentieth-century fossil experts saw ape-like and human-like characteristics in the specimen. Piltdown remains one of the lengthiest hoaxes in the history of science and it also remains unsolved – the perpetrator has never been successfully identified, although the fossil's discoverer, Charles Dawson, is the most likely suspect.

A short-lived counterpart to the Piltdown story occurred in May 1911, when the discovery of a 'prehistoric human skull' from the Devil's Cave near Steinau, east of Frankfurt, caught the attention of German anthropologists and the media. For years, building contractor Albert Lüders had wanted to turn the Devil's Cave into a tourist attraction. He had been drilling away to install the necessary infrastructure, and the skull was found near one of the drilling sites.

Over the following months the skull was discussed relentlessly in the media, with the German anthropologist Hermann Klaatsch championing its authenticity, and suggesting to German intelligentsia circles and newspapers that it might even have belonged to a 'neanderthaloid', or Neanderthal. Between mid-May and 10 August, roughly 50 articles on the topic appeared in 20 newspapers. Upon more rigorous investigation, anatomist Friedrich Heiderich revealed the skull to be nothing more than a chimpanzee skull that had been chemically treated with potassium permanganate – to make it look older – then stashed in the Devil's Cave. The fraudster, Wilhem Rappe, the village's pharmacist, whose brother had brought the skull back as a gift from his travels to Cameroon, was a notorious prankster in Steinau, and he promptly 'fessed up to the hoax. The German press had a field day raking Klaatsch and Lüders over the coals once their mistake was revealed. Unlike Piltdown, this 'fossil' didn't have any greater impact on the study of anthropology, but the case does demonstrate that authentication of new fossils has always depended highly on the provenience of their discovery, and has done throughout history.

There is a myriad of reasons why the Piltdown hoax (and its like) was so successful. However, one of the most intriguing reasons is a rather subtle one; it was a forged fossil, but it was made from real bones. Because of researchers' own preconceptions about evolution, anatomy and science, Piltdown was exactly what they were looking for – the 'missing link' between apes and humans they had predicted. Ninety years after the Piltdown hoax, another genuinely fake fossil flew through the scientific community. According to *National Geographic*, the story of this fake, *Archaeoraptor*, begins near the small village of Xiasanjiazi, in China's northern Liaoning province. Built out of genuine fossils and very real financial expectations, it reminds us that there is still a plethora of fossils that are less than authentic.

Ever since the 1920s, the Liaoning province has been a mecca for fossil hunters and palaeontologists. Due to the region's geology its fossils are well preserved, even including instances where dinosaur soft tissue has fossilised. In recent decades the area has become famous as a site that has produced the fossil evidence for the earliest flower,

some of the earliest mammals and the oldest intact embryo
of a pterosaur, in addition to more recent discoveries of
fossils of feathered dinosaurs. The region is one of the
world's most prolific and significant areas for discovering
fossils.

In July 1997, the story goes, a Chinese farmer found a
rather rare fossil – one of a toothed, bird-like creature that
had feather impressions. The farmer recognised that the
bigger and more complete the fossil was, the more he would
be paid for it. While academic palaeontologists often shun
the practice of buying fossils from local fossil hunters, it
does happen for the commercial fossil market. The farmer
glued together different parts of various fossils and the
result was spectacular. It sold.

In June 1998, having been smuggled out of China, the
fossil was sold to an anonymous dealer. By 1998, rumours
circulated through the palaeontological community that
a feathered dinosaur was in the hands of a private
collector, and in 1999 the fossil sold for $80,000 at the
Tucson Gem and Mineral Show to Stephen and Sylvia
Czerkas, owners of the Dinosaur Museum in Blanding,
Utah. (The sale was underwritten by the trustees of the
Dinosaur Museum so that the fossil would be available for
scientific analysis.) It was considered to be the first physical
evidence of feathers on a dinosaur, indicating a common
ancestor with birds.

However, the scientific community was sceptical about
the fossil, as amazing as the discovery was. There was
something ... 'off' about it. The fossil was refused publication
in prestigious peer-reviewed journals like *Science* and *Nature*,
as its provenience was particularly sketchy, and many
scientists refused to work with a specimen that ran counter
to international best practices for curating fossils. Concerns
were also raised about whether the fossil was a single
specimen or parts of multiple individuals.

In 1999, palaeontologist Tim Rowe of the University of Texas at Austin completed a detailed CT scan of the fossil as part of his work on the original team's scientific analysis. Rowe's work raised the possibility that the fossil was in reality a composite – not a single individual fossilised organism, but several stuck together. Scepticism be darned, was the Czerkas' response to Rowe's findings. The discovery was touted by *National Geographic* as a 'missing link' and was lavished with press coverage on its publication.

Further analysis of the CT scans indicated that *Archaeoraptor* was a chimera of a specimen, just as Rowe suspected. The fossil in question was actually five different individuals, from three different dinosaur species. The illegal exporters of the fossil had had a very clear sense of what would hit it big with audiences, even though many in the scientific community (including some of the original reviewers of the fossil) considered the possibility that it was a composite from the start. The fossil is regarded as a fake, although the species name *Archaeoraptor liaoningensis* remains in the scientific literature, much to the consternation of many palaeontologists who would like to simply rid themselves of the problematic taxon.

The *Archaeoraptor* fossil was a fake, to be sure, but it wasn't a fraud in the way that Piltdown or Beringer's Lying Stones were. It wasn't created to necessarily deceive its scientific audiences. Rather, it was glued together in the hopes of bringing in a better price to the discoverer, who was trying to eke out a living. (By some estimates, finding a fossil as spectacular as *Archaeoraptor* could result in a finder's fee that would be roughly equivalent to two years' salary. And subsequent excavations scientists actually did find feathered dinos in the area, bolstering the interpretation that *Archaeoraptor* offered.) While Piltdown is described as a hoax or a fraud, *Archaeoraptor* is most often described as a composite – certainly a more compassionate category in the taxonomy of fake fossils.

There are perfectly legitimate reasons to create composites; the problem arises when the composite has a tinge of deceit about it. The scientific world was looking for a feathered dinosaur at the turn of the twenty-first century, and a feathered dinosaur was exactly what it got.

★ ★ ★

Seeing what one wants to see is one thing. But being given what one expects is the market's side of the coin. With the burgeoning commercial market, which prizes complete fossil skulls, polished teeth and impeccably prepped specimens, there's a clear set of expectations from buyers about the fossils that their money buys, and international purveyors, especially from China, are only too happy to oblige. It's clear that people will pay exorbitantly for a spectacular trophy fossil, but not so much for scrappy bits and pieces, which results in the production of more Franken-fossils – those composite fossils that lead people to think that they're authentic. This is where the worlds of fossil smuggling, legislation, law enforcement, science and the commercial fossil market are coming to a head.

It turns out that Franken-fossils aren't as rare as one might think. Because the rise of the commercial dinosaur fossil market has created even more demand for them, the scientific community is constantly honing new methods of flushing out fakes. One of the most recent developments in rooting out fake fossils comes from Thomas Kaye of the Burke Museum in Seattle. In 2015, Kaye and eight of his international colleagues published a methodology based on Laser-Stimulated Fluorescence, or LSF for short. The authors pointed to a myriad of applications for the technique within the palaeontological community, not least of which would be an inexpensive way of sorting authentic fossils from bogus ones.

Their method is simple and, unlike CT scanning, doesn't require an immense investment of time or equipment. Their set-up involves a simple laser and a way of diffusing the light over the fossil in question. Since rocks, minerals and elements all fluoresce at different colours, when the different materials are excited by a laser beam, the fossil, rock and whatever else in there will light up as different colours, easily visible to the naked eye. 'Laser fluorescence opens the door to discovering previously unseen and unknown features in ancient fossils,' Kaye said in an interview with the Burke Museum.

The fossil parts of the rock are one colour; the resin, the plaster and the whatever else shine a different colour along the colour spectrum. With the help of a time-release digital SLR camera, images of the fluorescence pattern can be documented. The entire set-up costs something like $500 and offers an incredibly straightforward answer to the question, 'What am I looking at, really?' The answer will usually be: a fossil. But it's so much more than 'just' a fossil. You're seeing all of how of a fossil's gaps are filled in and what unscrupulous people use to do the filling. Burke's publication offers a dramatic 'before' and 'after' series of images; his team fluoresced and photographed a *Microraptor* specimen – a genus of dinosaur with wings and feathers. The glue and resin in the specimen lit up like a constellation. One would be hard pressed to interpret that *Microraptor* fossil as anything but a composite of materials as well as an inadvertent material record of decisions made about its conservation and preservation. While composed of not-real bits it is, nevertheless, an authentic fossil.

For many caught smuggling fossils, their own cases hinge on that ambiguity – on those bits of the fossils that aren't in fact real. They call these genuine fakes 'craft rocks', 'replicas' or 'artistic interpretations'. Phrases like these leave implicit the idea that the fossil in question is a creative endeavour, like a work of art under copyright, not an object in and of itself. Since the fossil is 'more' than just a fossil (it's had other fossil parts added to it, is a composite of multiple fossils and has had plaster and other things sculpted around it), it's no longer an excavated fossil but is an artistic work. This was one of the lines of defence that Eric Prokopi took in 2013 when he was charged with smuggling a massive *Tarbosaurus bataar* (sometimes called *Tyrannosaurus bataar*) fossil out of Mongolia via the United Kingdom and selling it on the US commercial market. The court was having none of the artistic license argument, however, and in *The United States v. One Tyrannosaurus Bataar Skeleton*, a federal judge convicted Prokopi of smuggling, sentencing him to three months in jail.

These sorts of Franken-fossils are a tricky bunch of fakes. While there were real bits and pieces of fossils in each specimen, the finished product, as it were, couldn't be scientifically named, because it wasn't, taxonomically, anything real. And this is certainly true historically – lots of fake things get names. Like Piltdown, they're made of genuine stuff, but they are fake animals. Like *Archaeoraptor*, they're assembled, composite specimens, made to sell well. And like Beringer's Lying Stones, they're sculpted works of art, carefully crafted for a purpose. They defy categorisation. Essentially, these are fossils that pander to the expectations of buyers as to what a perfect specimen 'ought' to look like. These Franken-fossils play to the market's expectations of real fossils and the legal possibilities of owning them. Although composites of fossils and casts are a longstanding tradition within the science and museum worlds of natural history, those composites don't attempt to deceive audiences, whereas Franken-fossils do.

Authenticity is irrelevant when giving audiences exactly what they didn't know they were asking for and exactly what they want. The commercial fossil market is whatever consumers want and, for many, what they actually want is something fake – something that can pass itself off as real. Piltdown and *Archaeoraptor* perfectly bookend the twentieth century's experience with fake fossils, and offer cautionary tales about seeing what we want to see. Popular media dubbed *Archaeoraptor* the 'Piltdown chicken'.

★ ★ ★

This brings us back to the story of Dr Johann Bartholomew Adam Beringer and his fake fossils. Even as early as the 1726 court proceedings, people had taken to calling Beringer's fossils *Lügensteine* – Lying Stones – and that's certainly how history has remembered them. Curiously, though, history has also created an odd sort of mythos around the story of Beringer and his fake fossils, suggesting that the entire episode was simply a rather sophomoric prank – perpetrated by his students, not his colleagues – that got out of hand. (Other tellings of the story relish casting Roderick and von Eckhart as villains, practically twirling their moustaches and capes with glee as Beringer makes a fool of himself.) For centuries, Beringer's story as the student-led prank has been told and retold, becoming a little less true every time – until the mid-twentieth century, when Beringer's court records were rediscovered in the Würzburg city archives in 1935.

The carvings themselves didn't last particularly long as genuine or authentic objects – and in all honesty, few other than Beringer counted them as such. But how they've dominated the Beringer narrative is certainly very real. It's a curious fake story that has evolved to explain fake fossils, perhaps to cover up Beringer's very real embarrassment. 'Like all great legends, this story [Beringer's Lying Stones] has a canonical form, replete with conventional moral messages, and told without any variation in content across the centuries,' essayist and palaeontologist Stephen Jay Gould argued in *The Lying Stones of Marrakech*. 'Moreover, this standard form bears little relationship to the actual course of events as best reconstructed from available evidence.'

Today, 433 of Beringer's specimens are still in existence, scattered in 14 museum collections in Germany, the Netherlands and Britain. (Oddly, 60 more of Beringer's Lying Stones have been lost during recent years, but are technically part of the historical record, thanks to photographs.) In 1835, two of Beringer's original Lying Stones were gifted to the eminent English geologist and palaeontologist William Buckland, who it would seem wasn't quite sure what to make of the gift. The specimens at the Oxford University Museum of Natural History show boomerang-like carvings on one of the mudstone pieces, and a slug-like creature with very cute antennae on the other. When the museum's curator pulled the two of the collections I could actually see the knife-carving marks along the edge of the slug. Truth be told, when I turned over the Lying Stones to get a better look, I burst out laughing.

Despite this, the Lying Stones have become the Real Thing outside of any scientific merit. Casts of the Lying Stones pop up in a plethora of museum collections. In 1931, the Natural History Museum in London borrowed 10 Beringer specimens and created casts for its own palaeontological collections; today the museum has more modern silicon moulds of the curios. The Natural History Museum uses the real casts of fake fossils for teaching purposes and to show off the historical curiosities of the palaeo collections.

In the 1970s, Phil Powell, then assistant curator for the geological collections of the Oxford University Museum of Natural History, created moulds and casts of the Buckland specimens for a Yorkshire children's television programme (*Extraordinary!*). In 1990, the British Museum ran an exhibition dedicated to the idea and question of 'fakes', and wrote to Mr Powell to enquire about the use of the Lying Stones in the exhibition. Mr Powell's letter of 22 November 1989 states, 'We have excellent casts of the originals. Would they be acceptable in an exhibition of fakes?' The British Museum, however, was hoping for more: 'If we may we would like to borrow the original stones. This is strictly an exhibition of genuine fakes!' (The real specimens were in fact included in the exhibition, which ran from 8 March to 3 September 1990.) By the latter half of the twentieth century, fossil enthusiasts, collectors and museums could buy casts of the Oxford stones through the Educational Palaeontological Reproductions Company in Essex. The 'fossils' live on.

'Beringer's story is not just an amusing anecdote in the annals of geology, nor should it be seen as a simple story of credulity told to unwary undergraduates. Using the modern definition of a fossil as the mineralized remains of a once-living organism, Beringer's figured stones look ridiculous, clearly the work of a third-rate sculptor,' historian of science Susannah Gibson argues, championing the need for context in Beringer's story. 'But by eighteenth-century definitions, Beringer's fossils could easily have been genuine ... The whole story turned on knowing the true nature of fossils.'

As for Beringer, it's important, of course, to not judge him by the standards of today, although even by the standards of 1725 he failed spectacularly. Recall that Beringer opted not to confer with his scientific community and ignored the heavy hints about the stones' authenticity

CHAPTER THREE

Carbon Copy

On 25 April 1772, the French chemist Antoine Lavoisier set fire to a diamond. Several diamonds, actually, and on purpose.

Lavoisier and his colleagues, the chemist Pierre Macquer and the apothecary Louis Claude Cadet, burned those diamonds by putting them into a crucible, then placing the crucible in the furnace in Cadet's workshop. The experiment was, nominally, about determining whether air was necessary to burn a diamond, or if the gem could burn in a vacuum. The trio proceeded to put a plethora of other gems in the furnace, much to the shock and awe of members of their informal audience, such as Paris's eminent Duc de Croÿ, a fixture in popular intelligentsia circles of France, who had stopped by Cadet's shop that afternoon.

'I found the furnaces lit and everyone very busy. Monsieur Lavoisier, Farmer General and Academy member, and somebody else were trying to distill a diamond,' the Duc de Croÿ wrote, his shock and disbelief unmistakable. 'These experiments are putting lots of diamonds in the fire!'

These were not the first diamonds to be put 'in the fire', however. In 1694, Giuseppe Averani and Cipriano Targioni, two Florentine naturalists from the court of Cosimo III, used 'burning glasses' – large lenses of biconvex glass – to focus rays from the sun on to diamonds that were promptly converted into an ashy substance known as a calx. Isaac Newton incinerated diamonds as part of his various research experiments. And in 1760, the Austrian Emperor François I underwrote a wildly expensive experiment where 6,000 florins' worth of diamonds and rubies were enclosed in cone-shaped crucibles and stuck in a furnace for 24 hours.

'The rubies had changed,' the courts' experimenters reported, 'but the diamonds had completely disappeared, to the point that not the slightest trace was found.'

By the mid-eighteenth century, chemists throughout Europe were interested in observing, defining and documenting the physical properties of gems, minerals and elements. Diamonds were no exception and the work of these previous experimentalists provoked further curiosity. Lavoisier wanted to understand how the chemistry of combustion worked in diamonds – why the hardest mineral in the world could be changed from a shiny, brilliant gem into charcoal, and why this didn't happen with other gemstones.

Five days after the Duc de Cröy's visit, on 29 April 1772, Lavoisier reported his results from the tests on 21 diamonds to the Academy of Sciences, noting that there was a difference in a diamond's ash when it was exposed to air and when it was not. But these results posed more questions than they answered. What was happening to those diamonds? Were they simply vaporised, as happens to water when it is heated? Were the diamonds actually combusting? Or were they simply breaking down into their most basic particles 'so fine that they can no longer be perceived by our senses'? More experiments, Lavoisier concluded, were necessary. As were more diamonds.

And, on 8 August 1772, more experiments and more diamonds were what Lavoisier got. Instead of using a furnace, Lavoisier and his colleagues – Cadet, the natural philosopher Mathurin Jacques Brisson and the pharmacist Pierre François Mitouard – opted for a different experimental set-up.

Lavoisier ordered a large, six-wheeled wooden platform wagon with enough space for eight or nine scientists to work. He had two massive burning glasses – convex lenses – mounted on one end of the platform that could be manipulated through a complex series of cogwheels and cranks to best capture the sun's rays as the sun moved across the sky throughout the day. The lenses were 2.4m (8ft) in diameter, jointed at the edges, and filled with 79.5 litres (140 pints) of alcohol as a way to make the lenses even thicker, intensifying and amplifying the sun's rays. (A 'liquor lens', per Lavoisier's notes.) Lavoisier outlined a programme of experiments and organised all the necessary materials.

'A group of assistants were in charge of adjusting and moving the machine while the four scientists, wearing wigs and dark glasses, proceed with the test, installed as if they were on the poop deck of a ship,' historian of science Jean-Pierre Poirier explains in his biography of Lavoisier. Because the experiment was set up on the vast south-facing terrace of the Jardin de l'Infante, which stretched between the royal palace and the banks of the Seine, 'Elegant women and the curious came to stroll there and observed the scene with amazement.'

Between 14 August and 13 October 1772, Lavoisier conducted 190 experiments on diamonds, metals and other minerals with this set-up. Then, between 14 March and 14 August 1773, he carried out an additional set of 19 experiments on diamonds – in glass jars – to be ignited by the sun's rays. (Unfortunately, the glass jars broke as a result of the incredibly high temperatures.) What Lavoisier and his colleagues were left with in every instance of their diamond experiments were piles of charcoal – in amounts that weighed the same as the diamonds that they had started with. Based on these experiments, as well as a number of others conducted over the next several years, Lavoisier concluded that the charcoal and diamonds were simply forms of the same material. When he published an updated list of chemical elements in 1789, he referred to this substance as carbon.

In the 1790s, the English chemist Smithson Tennant expanded Lavoisier's work with diamonds, and through his own sets of experiments, demonstrated that the gem is nearly pure carbon. (In his report to the Royal Society of London, Tennant noted that, 'Though [Lavoisier] observed the resemblance of charcoal to the diamond, yet he thought that nothing more could be reasonably deduced from their analogy, than that each of those substances belonged to the class of inflammable bodies.') Tennant converted identical weights of charcoal and diamond to exactly the same volume of carbon dioxide gas, establishing that charcoal and diamonds were chemically equivalent when they were both in solid form. 'As the nature of the diamond is so extremely singular,' Tennant reported to the Royal Society in 1797, 'that it consists entirely of charcoal, different from the usual state of that substance only by its crystallized form.'

Understanding what a natural diamond was made of was the first step towards thinking about how to reproduce it in a laboratory. The first reports of a 'man-made' diamond came just over 100 years after Lavoisier and Tennant demonstrated that diamonds were made of carbon. Their experiments opened the door to the possibility of transforming carbon – like its very common form of graphite – into diamonds. While forgers and fraudsters had been passing off fake diamonds for millennia, the idea that a diamond could be replicated – that it could be synthesised and still be 'real' – caught the imagination of many nineteenth- and early twentieth-century experimental scientists.

And thus begins the story of how to make a real, non-natural diamond.

★ ★ ★

On 19 February 1880, Professor Mervyn Herbert Nevil Story-Maskelyne was working as Keeper of Minerals at the

British Museum, where he received a small package of nine crystalline specimens. (This Story-Maskelyne was the grandson of the royal astronomer with the same name, famous for devising a way to mark longitude while at sea.) James Ballantyne Hannay, a 25-year-old Scottish chemist who claimed that the pieces were diamonds that he had made himself, had sent the crystals to him.

Professor Story-Maskelyne was an expert in gems, mineralogy and chemistry, having taught at Oxford for decades. After examination and a series of stringent tests, Story-Maskelyne authenticated the diamonds as real – albeit 'man-made' – and proceeded to report Hannay's incredible achievements in *The Times*, as well as submitting a report to the scientific intelligentsia of the Royal Society of London.

James Ballantyne Hannay was a practising chemist and metallurgist. In his numerous experiments, he had observed that many substances (like silicon, aluminium and zinc) were insoluble in water at normal temperatures and would largely dissolve in water vapour at high temperatures. Hannay thought such a solvent might be found for carbon that would allow the structure of a chunk of carbon to rearrange itself from one form into another. In short, Hannay wanted to create a dissolving material in an appropriate solvent, and then figure out how to induce crystalline growth in a particular pattern – not unlike the process of making rock candy that uses a seed crystal to start the growing pattern. 'I thought that if I could by increased temperature dissolve the nascent carbon in the metal I might obtain a diamond,' Hannay reported to a meeting of the Royal Society in April 1880.

However, obtaining the diamonds that Hannay sent Story-Maskelyne wasn't easy or straightforward, because designing and building an experimental set-up capable of reaching the high heat and pressure necessary was, in a word, difficult. By September 1879, Hannay had started using wrought-iron tubes that were 50cm (20in) long and 1.5cm (1in) thick, with a 1.2cm (.5in) bore hole in the middle – each tube was roughly the size of a baton. The tubes acted as crucibles to isolate carbon, facilitating its transformation into a diamond. Hannay filled each two-thirds full of a 'paraffin spirit' and bone-oil mixture, adding 4g of lithium, and a pinch of lampblack for the carbon-base solvent. He heated the tubes to a 'dull red-heat' for 14 hours, then set them aside to cool.

Incidentally, finding a successful way to seal the tubes was not an insignificant part of Hannay's experiments. The heat and pressure they were subjected to meant that they wouldn't just stay screwed shut – they had to be crimped or welded. But when Hannay crimped end plugs – balls with an appropriate diameter – to the tubes, this inadvertently and invariably turned the tubes into miniature cannons. Once the pressure inside built up after hours in the furnace, the end of the tube would shoot out. 'It was expected that the pressure would only serve to make the closing more secure, but, on heating, the iron yielded and the ball was driven out with a loud explosion,' Hannay clinically noted in his report.

Nine out of ten times, the iron tubes exploded, destroying the furnaces and their contents, and essentially rendering Hannay's workplace something akin to an active battlefield, repeatedly leaving his laboratory in shambles. 'The experiments are, however, too few, and the evidence too vague, to draw any conclusions, as there are even very few negative experiments from which anything can be learned, most of the results being lost by explosion,' he acknowledged.

Of the few tubes that survived Hannay's ministrations, most of the carbon inside either 'evaporated' out completely or appeared 'soft and scaly', as he reported to the Royal Society. But a precious handful – three tubes out of

80 runs – had crystals inside that were hard and transparent. Under a microscope they looked like diamonds, and these hard, transparent, diamond-like specimens were what Hannay sent to Professor Story-Maskelyne in February 1880 and what Story-Maskelyne authenticated as real diamonds. It would seem that Hannay really had made the world's first genuine synthetic diamond.

Although Hannay's results resonated with some prominent scientists, like the Scottish chemist Sir Robert Robertson – the first scientist to establish two types of natural diamond – the scientific community politely but pointedly ignored Hannay, his claims and his diamonds, despite the backing of Robertson and Story-Maskelyne. (Fifty years before Hannay sent his diamonds to Story-Maskelyne, the British publication *Mechanics' Magazine: Museum, Register, Journal, and Gazette* printed a letter summarising a meeting of the French Academy of Sciences, where a gentleman named Monsieur Gannal claimed to have successfully 'converted pure carbon' into 'crystals possessing all of the properties of the diamond'. But that claim was never taken seriously.) Hannay died in 1931, having moved on from the project of making diamonds in his later years, but remained utterly convinced that he had succeeded.

Scientific and historical communities have never been exactly sure what to make of Hannay's crystals. For over 100 years, opinion has shifted backwards and forwards about whether the specimens that Hannay created were in fact diamonds and, if they were, whether Hannay actually created diamonds in his lab or if the specimens he sent to Story-Maskelyne were naturally produced diamonds that he attempted to pass off as his own handiwork. (In 1902, Hannay quietly refuted a suggestion made by the *Encyclopaedia Britannica* that his diamonds were simply carborundum or silicon carbide.) Throughout the twentieth century, experts have examined the nine specimens that are still at the British Museum, testing them against increasingly sophisticated methodologies that have developed since Story-Maskelyne's original assessment of the stones.

More methods and more tests, however, have not made the Hannay story any more straightforward. The eminent crystallographer Professor Kathleen Lonsdale of University College London, who herself worked on the question of synthesising diamonds, examined Hannay's diamonds several times over many decades, eventually concluding that the gems were natural diamonds. Further examination of the diamonds in 1968 and 1975, using a series of tests that were capable of differentiating laboratory diamonds from naturally grown ones, concurred with Lonsdale's assessment that the diamonds were, indeed, diamonds, but that they were natural – Hannay hadn't made them.

All of these mid-twentieth-century reports charitably concluded that the natural diamonds had been 'contaminants' in Hannay's studies – natural seed diamonds that Hannay used to encourage crystal growth the way synthetic diamonds are made today, although Hannay never noted that he had used seed diamonds in his meticulous lab records. Some recent historians do not buy the contaminant theory and contend that Hannay was an out-and-out fraud, while others argue that Hannay's colleagues and workers seeded the tubes with natural diamonds in the hopes of staving off more dangerous experiments.

What is indisputable, however, is that the question of making a synthetic diamond certainly caught the public and scientific imaginations of the nineteenth century. While the scientific community might have, more or less, ignored Hannay's efforts when he first published his results, the question of whether a laboratory could reproduce a mineral made by nature was of utmost interest among experimentalists, as diamonds became better known as a result of colonial mining and trade. This also neatly intersected with then-current research interests to do with materials science.

In 1893, following Hannay's efforts, the French experimental chemist Ferdinand Frédéric Henri Moissan tried using his newly developed electric-arc furnace to transform graphite into diamonds. Moissan heated iron with sugar charcoal in a carbon crucible to 4000°C (7232°F) in the laboratory furnace, then plunged the white-hot crucible into cold water, solidifying the iron and thus exerting a very high pressure on the carbon. These experiments led Moissan to the discovery of 'moissanite', a silicon carbide, which he initially mistook for diamond, due to the mineral's hardness.

Moissan's experiments attempting to create diamonds were so well known that in 1904 his fellow Frenchman, Monsieur Henri Lemoine, exploited Moissan's celebrity, claiming to have been able to reproduce Moissan's laboratory diamonds and wanting to set up a laboratory-factory funded by wealthy investors. (Historical lore has it that Lemoine shocked his potential investors by performing a demonstration of a diamond-making experiment *au naturel* to prove to his audience that the technology to make diamonds was feasible and that he hadn't simply hidden natural diamonds in his clothes.) Over the next three years, Lemoine bilked investors out of 64,000 pounds sterling, an enormous sum of money. With the 1908 confession of a jeweller who admitted selling Lemoine a cache of small, uncut diamonds that matched the description of the gems Lemoine 'made' in his laboratory 'experiment', it quickly became clear that Lemoine's created gems were simply natural, river-washed diamonds from the Jagersfontein mine in South Africa. Lemoine was subsequently arrested, tried for fraud and found guilty. The scientific and historical juries are still out as to whether

what Moissan created were actually diamonds, since no
one was able to reproduce his results.

In the ensuing decades of the early 1900s, several
prominent scientists attempted to re-create both Hannay's
and Moissan's experiments – including Sir Charles Algernon
Parsons, inventor of the steam turbine. Many of these
efforts produced small, diamond-like specimens. Parsons
spent decades and a considerable chunk of his personal
fortune in his quest, but in 1928 he renounced his claims,
having become convinced that his synthesised gems
were not, mineralogically, diamonds. The fervour that
surrounded the manufacture of diamonds even crept into
the budding world of nineteenth-century science fiction,
when famous novelist H. G. Wells published a short story
called 'The Diamond Maker' in 1894, based on Moissan's
experiments. However intriguing these early attempts
were, none of the experiments created reproducible results.
And none yielded 'real' synthetic diamonds that were
universally accepted by the scientific community.

★ ★ ★

'Some gemstone experts are apprehensive about synthetic
gemstones,' chemist and mineralogist Kurt Nassau offered
in the introduction to his 1980 book, *Gems Made By Man*,
as tensions built in the gemmological world about how to
make sense of non-natural diamonds. 'They regard them as
intruders to be shunned.'

The suspicion of non-natural gems runs deep because
for most of history, the only kinds of gem that didn't
come from nature were fakes. Consequently, fake gems –
fake diamonds – are nothing new. Like frauds and forgeries
that we find in art, manuscripts, codices and fossils, fake
gems find a plethora of ways to creep into the world of
genuine things, and have done for millennia. Fakes serve as

a narrative foil to synthetic diamonds, and the relationship between synthetic, natural, fake and real is conflated and contorted. This is particularly true in the twenty-first century, when synthetic diamonds constantly come up against thousands of years of cultural mores that might see them as less than real.

In the first century AD, Roman author and natural philosopher Pliny the Elder took on the question of such fakes in the mineralogy section of his famous *Historia Naturalis*. Among the descriptions of various gems and minerals (where he reports that quartz is a form of ice produced by the congealing of water in the extreme cold), Pliny calls his reader's attention to the proliferation of fake gems – specifically, instances where fraudsters simply substituted cheap look-alikes for the genuine thing. He attributes the abundance of fakes to humanity's obsession with precious gems – his disdain is palpable: 'there is no other kind of fraud practiced, by which larger profits are made'. In order to combat the rampant gem deception across the Classical world, Pliny offers the first description of a scratch test to differentiate real diamonds from fakes – noting that real diamonds would scratch other minerals but not vice versa.

Most fake diamonds were attempts to pass off pieces of glass or quartz, with the hope that the buyer would be gullible enough to not notice a substitution. As very few diamonds were cut or faceted in ancient Rome, such substitutions were rather straightforward. Other fraudsters turned to alchemy as a clever way to dye stones to look like more expensive gems. The Stockholm Papyrus, for example, was a collection of alchemic recipes written in Greek in AD 200–300, which contained 71 how-tos for creating fake gems. It walked its readers through how to take selenite, topaz or moonstones and colour them to look like emeralds, rubies or beryls. (In going through those

Stockholm Papyrus alchemical recipes, reader, I was struck by the number that called for the use of 'tortoise bile'.)

Yet these were all simply ways to approximate but never replicate nature. 'For although art may imitate nature nevertheless is cannot reach the full perfection of nature,' wrote Albertus Magnus, the thirteenth-century Dominican bishop and natural philosopher, in his *Book of Minerals*. Magnus was talking about glass – it could look exquisite and sparkle like a diamond, but only as an imitation. It was not natural. It could never be a 'real' gem.

Which brings us to the differences of intent between those creating and peddling fake diamonds, and what was going on with Moissan and Hannay's experiments. Compound the differences of intent with the ever-shifting definition of imitation, and it's no wonder that just what gets to count as a 'real' diamond is shaky. Synthetic gems are not merely imitations (simulation, a fake, a paste, a glass) of natural gemstones. To deserve – to earn, really – their designation as synthetic replicas they have to have the appearance, chemical composition, crystal structure, hardness and optical properties of a natural stone.

This is a huge change in the story of non-natural diamonds. In previous centuries, human-made diamonds were inherently frauds and hoaxes – fakes. With the attempts to transform one form of carbon into another – a piece of graphite into a diamond – the intent and context of the not-natural diamond changed. Was it possible, scientists asked themselves, to produce in their laboratories what nature had made millions of years ago, without the deception that had always surrounded fake diamonds?

So, when scientists began growing, manufacturing and synthesising diamonds, the research grew out of Lavoisier's combustion experiments, rather than the fraudsters' shilling glass imitations. (Lavoisier was very careful to position himself as a proper *chemist* – not an alchemist – to lend more legitimacy to his research, distancing his work from the older discipline's reputation for fraud and deception.) The story of made diamonds is not the story of a fake becoming real, in the way Spanish Forger art pieces gained value in the art market. Synthetic diamonds were real diamonds to begin with, comparable and identical to natural diamonds on an elemental level. This shifted the story from fraud into one of scientific ingenuity – only the story of their origins separates the two.

★ ★ ★

To that end, the first non-natural diamonds were made in General Electric's laboratories in December 1954.

Ever since Lavoisier and Tennent's experiments, scientists and engineers had known that they needed to subject carbon to enormous amounts of heat and pressure in order to transform it into a diamond, but just how to pull this off required no small amount of trial and error. As demonstrated in the experiments of Hannay, Moissan and even Parsons, creating immense amounts of pressure is difficult, if not downright dangerous. But, their early experimental designs weren't necessarily flawed – they just needed different technologies and manufacturing to be able to create those pressures in the lab. The answer came from the American Nobel-winning physicist Percy Bridgman, who worked with the development of a vertical hydraulic piston that pressed into a cylinder, which created 4,200 atmospheres of pressure through a complex anvil system. For years, Bridgman's laboratory at Harvard had holes in the walls where the canister – affectionately known as 'The Bomb' – blew out and embedded materials in the walls. Bridgman, it was reported, superstitiously never had the holes repaired.

By the 1940s, the General Electric Research Laboratory in Schenectady, New York, had become the centre of synthetic diamond research, bringing together researchers from chemistry, physics and industrial engineering. (Schenectady had a long tradition of supporting speculative research projects even if they weren't directly related to the production of electrical equipment.) The team was made up of Francis Bund, Herbert Strong, Howard Tracy Hall (generally H. Tracy Hall or Tracy Hall in popular literature), Robert Wentorf and James Cheney, and was managed by Anthony Nerad. The project was code-named 'Project Superpressure' and everyone was sworn to secrecy. Drawing on Bridgman's work, Superpressure used several different apparatuses in its experiments. For years, the team devoted extraordinary time, effort and resources to making synthetic diamonds – and, more importantly, to learning how to manufacture them in a way that was replicable. As the years went by, the management at General Electric began to worry that manufacturing diamonds would turn out to be gimmicky at best and a total monetary sinkhole at worst. By December 1954, the team needed concrete, tangible results – diamonds – to justify its work.

On the evening of 8 December 1954, Herbert Strong began Experiment 151, setting the pressure cone apparatus at an estimated 50,000 atmospheres, cranking the temperature up to 1250°C (2282°F), and depositing a carbon and iron mixture with two small natural diamonds to seed diamond crystal growth. It was not unlike the methods used by Hannay decades earlier, only Strong was clearly using seed crystals. (Research in the Soviet Union used seed diamonds as part of the effort to grow diamonds, as near as General Electric could tell.) Most of Strong's earlier experiment runs had been short, a couple hours at most. But this time he decided to let Experiment 151 mimic nature – which took millions of years to produce diamonds – and, at least, to extend the time of the experiment and let it run overnight.

On the morning of 9 December, the two seed crystals tumbled out freely, unchanged in the crucible. A blob of the iron-carbon mixture had melted into one end of the tube and Strong sent the blob to the metallurgy division to be polished. A bit miffed, metallurgy sent back a message on 15 December, informing Strong that they were unable to polish his sample because it was destroying the polishing wheel. Whatever was in the blob was strong and hard – hard enough to gouge metallurgy's equipment – and only a diamond could be that tough. Strong recounts, 'The entire group gathered around to inspect the hard point. Initially there was a moment of stunned silence. Could it possibly be diamond? Finally, Hall spoke the verdict: "It must be diamond!"' Subsequent X-ray analysis confirmed that the diamonds in question were, in fact, laboratory made.

On 16 December 1954, Hall performed a similar experiment by himself, using an older bit of technology, a high-pressure press called a belt. He added two diamond seed crystals to iron sulphide and placed everything in a cylindrical graphite heater. Carefully following the belt protocol he had designed after months of working with the apparatus, he placed thin disks of tantalum metal between the sample and the belt anvils to facilitate current to heat the sample. Everything was cooked at 1600°C (2912°F) under 100,000 atmospheres of pressure. The entire experiment took 38 minutes.

'I broke open a sample cell after removing it from the Belt. It cleaved near that tantalum disk,' Hall said of his discovery, after seeing flashes of light from octahedral crystals that were stuck to the disk. 'Instantly, my hands began to tremble. My heart beat wildly. My knees weakened and no longer gave support. Indescribable emotion overcame me and I had to find a place to sit down!' In Hall's mind, there was no doubt about the results. 'I knew that diamonds had finally been made by man.'

Suddenly, after years of research and in the space of a mere week, General Electric had two possible ways to manufacture diamonds. In the weeks that followed the question wasn't so much whether researchers could make diamonds. The question was whether they could make them again according to either Strong's or Hall's experiment designs. Which method was better?

Researchers spent weeks trying to duplicate Strong's results and never could. (Strong contended that the heat had fluctuated significantly during the night of Experiment 151 and that fluctuation played a role in the run's success – serendipity as its finest.) Hall, working with Robert Wentorf, verified his original results rather decisively. Over the next two weeks the two of them successfully made diamonds 20 times, using Hall's 400-ton press and belt system. On 31 December 1953, General Electric had physicist Hugh Woodbury independently confirm Hall's diamond-making methodology.

Like many discoveries in the history of science, pinpointing exactly who ought to be credited with a discovery – and who history has credited – is a bit tricky, and the story of who manufactured the first synthetic diamond and when is no exception. In a series of publications, Strong has highlighted the work that the group did, pointing to the complicated nature of the problem and emphasising that the work was beyond what one man could claim to do.

Hall, on the other hand, felt ostracised from the team. (As a practising member of The Church of Jesus Christ of Latter-day Saints, he claimed to have been on the receiving end of religious prejudice during his tenure at General Electric.) He also felt underappreciated by the company: General Electric increased his salary from $10,000 to a mere $11,000 between 1953 to 1954, and paid him a $10 savings bond, despite making millions from his work. (This was somewhat typical for a Cold War research laboratory. Corporate scientists signed over the rights of their intellectual property to their parent companies and often received small bonuses like this one in connection with any patents that resulted from their work.) Hall left General Electric in mid-1955 to take a faculty research position at Brigham Young University, and authored several patents related to the manufacture of synthetic, laboratory-grown diamonds. He also started the company MegaDiamond, which eventually became General Electric's biggest domestic competitor in the diamond-making business. Both Hall and Strong have claims to be 'the first' to create diamonds, although most tellings of the story credit Hall because of the replicability of his experiments.

General Electric published the results of its laboratory-grown (or 'man-made' or synthetic in the parlance of the mid-twentieth century) diamonds on 15 February 1955. Reporters were invited to check the laboratory-made diamond dust under a microscope, and the research team was under strict instructions to keep mum about details of its work. Between February and March 1955, newspapers across the country blurbed General Electric's success, but were short on technical details for their readers. Most of those quoted in articles were jewellery experts, who dismissed these diamonds as any sort of financial challenge to the diamond market at that point. In the following months, General Electric held several more press events (one, for example, in May 1955, was at the Sheraton Hotel in Rochester, New York), which talked up not only the engineering prowess of its synthesised diamonds, but also how the project would be a 'boon to US industry'.

Superpressure's research didn't stop with press releases. Just because General Electric had a reliable method of producing diamonds, the logic went, it didn't mean that there wasn't plenty to explore in the world of synthetic diamonds. General Electric also wasn't so naïve as to think that others weren't pursuing the same goal. In fact, more than a decade earlier, in the early 1940s, Sweden's major electrical company, ASEA, had begun its attempts under the direction of the Swedish scientist and inventor Baltzar von Platen. ASEA ran extravagant experiments, and despite a series of complications not unlike those that General Electric faced, actually produced synthetic diamonds (diamond grit, technically, similar to the dust that General Electric showed reporters in that first press conference) in 1953, before team Superpressure produced theirs.

However, ASEA never published anything about its project. 'The most maddening, inexplicable aspect of ASEA's diamond-making victory was their absolute, absurd silence,' Robert Hazen laments in his book *The Diamond Makers*. 'After hundreds of years of concerted effort in which brilliant scientist after brilliant scientist had failed ... ASEA had triumphed.' In fact, it wasn't until two years later, after the Americans at General Electric had announced their own success, that ASEA deigned to make a brief statement about its experiments, publishing the work in 1960. In subsequent decades ASEA scientists explained that they felt that publishing in 1953 would have been premature, and that they wanted to have something more substantive when they did publish – they also claimed that they had no idea that other scientists were working on the same question.

Back in America, however, the race was on to patent the technologies and to corner the manufacturing market for synthetic diamonds. The US Army's Electronics Research and Development Laboratory at Fort Monmouth, New Jersey, became interested in pursuing the research question, in large part because diamonds had so many industrial applications, as did a research group at the University of Michigan. When scientists from the US Army met with General Electric researchers, all of the company's equipment was covered with paper drapes for secrecy.

In 1957, General Electric began selling industrial-grade diamonds, and in 1959 the team published the details of its discovery in *Nature* and filed a globally recognised patent for its 'Man-Made' diamonds, as rumours swirled that South African and Soviet labs were close to producing diamonds. In February 1960, Hall published a detailed description of a similar belt apparatus to General Electric's and filed his own patent, followed by two more over the ensuing years. The process of making a synthetic, laboratory-grown diamond was beginning to enter the competitive intersection of science and industry, and within a year historical estimates suggest that more than two dozen research groups had successfully synthesised diamonds.

Consequently, questions about how to make the laboratory process better, faster, cheaper and more reliable began to creep into the research, now that the proof of the concept was thoroughly demonstrated. The key to better diamond synthesis, Superpressure found, was to reach temperature and pressure conditions that were conducive to forming a diamond at the same time that the metal mixture was in a liquid state. Liquid metal – as hypothesised and used seven decades earlier by Hannay and Moissan – was necessary to dissolve the carbon source, in order to provide a steady supply of carbon atoms and to catalyse diamond growth. The high-pressure/high-temperature experiments produced a lot of other successes, in addition to actual diamonds, by creating new substances. The team began to look for new milestones.

And the team wasn't without a sense of humour. In December 1955, Robert Wentorf went to the local food co-op in Niskayuna, New York, bought his favourite type of crunchy peanut butter and brought it back to the diamond lab at General Electric. With a certain theatrical flair, he scooped out a spoonful of the crunchy peanut butter and ran it through the Superpressure's experiment protocols, transforming it into tiny crystals of diamond – thus demonstrating that a carbon base of any sort could be turned into a diamond, given enough heat and pressure.

★ ★ ★

In nature, diamonds start out as carbon-containing fluids, hundreds of kilometres below the Earth's surface, typically involving the breakdown of carbon dioxide or methane. While some diamonds – microdiamonds or nanodiamonds – form when a meteor hits the Earth, the majority of diamonds form deep inside the Earth's mantle, under immense heat and pressure, until they are transformed into diamonds that are carried to the crust.

The growth pattern of diamonds indicates that they grow slowly and are anywhere from three billion to a few hundred million years old. Although diamonds take eons to form, they reach the surface rather quickly – some scientific estimates put diamonds' travel time at months or even hours. Most diamonds come to the surface through a specific type of molten rock, called kimberlite, which works its way up from the mantle through vertical fractures in the Earth's surface. The kimberlite-magma pipeline is what we know as a diamond vein or diamond mine. Creating diamonds in a laboratory shortcuts millions, if not billions, of years of geologic time.

The 1960s saw the rise of jewellery-quality non-natural diamonds, both in size and in clarity, as the team at General Electric started producing diamonds that were at least a carat in size, at the rate of about one carat per week. Photos from the late 1960s to the early 1970s show researchers and General Electric executives posing with large, discernibly cut diamonds. These diamonds, the pictures subtly suggest, are 'real' diamonds – they are cut gems just waiting for their jewellery settings. Creating this sort of narrative around the synthetic diamonds translated their real-ness to sceptical audiences. The diamonds *looked* like gems, ergo they *were* gems, in a way that synthesising grit for industrial tools wasn't. In 1968, General Electric transferred the diamond project to Worthington, Ohio, where it remains today, known as the 'GE Specialty Materials Department'. By the late twentieth century, Sumitomo Electric Industries Ltd in Japan, De Beers Industrial Diamond Division (Pty) Ltd in South Africa, General Electric in the United States and Russian researchers in Novosibirsk were all creating diamonds that were going to either industrial or jewellery markets.

While the mid-twentieth century laboratory diamond research pushed towards new goals, all of the experiments and manufacturing depended on the technique of creating diamonds using high pressure and high temperature – what is known in the industry as HPHT diamond manufacture. (Even in the twenty-first century, HPHT is used to make most synthetic diamonds.) But in the late twentieth century, a new generation of diamond makers hypothesised that diamond layers could form from hot, vaporised carbon atoms at low pressure. This sort of method would, potentially, allow industries to manufacture synthetic diamonds on a much larger scale.

This new method, called chemical vapour deposition (CVD), was a recycling of experiments from the 1950s, but with the technology and engineering of the late twentieth century. In 1952, William Eversole of the US-based Union Carbide Corporation demonstrated that it was possible to grow diamonds at lower pressures if one started with a carbon-containing gas. Basically, CVD works by forming a gas of single, isolated carbon atoms from a hydrocarbon mix (heating carbon atoms to an incredibly high temperature), then encouraging the atoms to cool into a crystalline lattice structure, forming a film, much like how layers are formed on a 3D printer. Researchers began successfully creating greyish diamonds through CVD in the late 1980s to the early 1990s.

This method enabled more diamonds to be made over a larger variety of substrates and at lower temperatures and pressures. CVD for diamonds overlapped quite a bit with CVD research in semiconductors and wafers. The most common material used in semiconductors is silicone – silicone oxide – and the explosion of semiconductors for electronic chips and integrated circuits saw an industry refine and hone the technology necessary to be able to successfully use CVD techniques. Advances in CVD diamonds continue today, and although HPHT is still used for the manufacture of the majority of diamonds, CVD is quickly catching up.

★ ★ ★

The story of a natural diamond begins, of course, millions if not billions of years ago and deep within the Earth. Even once a diamond comes to the Earth's crust, it's still a long way from becoming the gem that sparkles in its setting, or even, for that matter, from becoming the diamond grit for industrial uses. In order for a diamond to become, well, a diamond, it undergoes intense cultural forces that morph it into the object that we know it as today. Diamonds are just as much the result of their history as their chemistry and geology.

For millennia, humans have manipulated diamonds – for jewellery, yes, but also for technology. In late Neolithic China some 4,500 years ago, for example, when craftspeople polished ceremonial burial axes made of the extremely hard mineral corundum, they used diamonds to obtain the exceptionally smooth finish. For thousands of years, peoples and cultures have exploited the hardness of diamonds in a variety of forms – from tipped tools to industrial grit – taking advantage of the mineral's physical properties. While Pliny the Elder described both the natural and cultural elements of diamonds (as well as the proliferation of fakes) in *Natural History*, cultural meaning beyond strict financial valuation varied greatly throughout human history over the last two millennia. The Middle Ages and the early Renaissance saw a plethora of meanings attached to the mineral. The eleventh-century poet Marbodus of Brittany, for example, suggested that diamonds were magic stones of great power that could be used to drive away nocturnal spectres, and he recommended wearing diamonds set in gold on one's left arm. In the twelfth century, Saint Hildegard claimed that the diamond allowed its wearers to ward off Satan, thus resisting his power day and night.

With the publication of his book *Travels* in 1360, the English writer and knight Sir John Mandeville proposed that diamonds could prove guilt or innocence when one was accused of a crime – if the accused was guilty, the diamond would dim and if innocent, it would sparkle and shine even more than it did before. (Mandeville also hypothesised that 'A diamond is synthesized when two larger ones, one male and one female, come together, in the hills where the gold is. And the diamond grows larger in the dew of a May morning.') For several centuries, diamond dust was considered to be a deadly poison – Florentine sculptor and goldsmith Benvenuto Cellini was convinced he was being poisoned by diamond dust when he was imprisoned on false charges in 1538. In the world of sixteenth-century natural philosophy, Italian Gerolamo Cardano claimed that diamonds make their wearers unhappy by preying frequently on the mind – not unlike the irritation of the sun being constantly in one's eyes. Since the stone was hard and transparent it was often ascribed moral characteristics, like invincibility and purity, which would rub off on to its wearer.

Throughout their history, natural diamonds have been material as well as metaphor, a duality of nature and culture that continues to shape what we think of as a real, authentic diamond even today. However, the cultural cachet of diamonds accelerated exponentially over the last 130 years due to De Beers and its diamond legacy. Indeed, it is impossible to talk about the cultural invention of diamonds – synthetic or natural – without talking about De Beers.

Until the late nineteenth century, diamonds were found primarily in India and Brazil. Historical estimates suggest that for centuries the entire production of gem diamonds could be measured by only a few pounds per year. When huge diamond mines were discovered in northern South Africa in the 1870s, near the Orange River, diamonds were 'being scooped out by the ton'. The British financiers who had backed the South African mines were concerned about what the influx of gems would mean – as the market at this time depended entirely on the gems' scarcity – and became convinced that if all of the mined diamonds went to market, diamonds would become semi-precious stones at best. Major investors moved to consolidate their diamond-mining investments to control production and 'perpetuate the illusion of scarcity' that shaped the diamond market.

And thus, in 1888, De Beers Consolidated Mines, Ltd. incorporated in South Africa. (De Beers's first president was Cecil Rhodes, an ardent and unapologetic British imperialist. Rhodes would go on to oversee the establishment of Rhodesia – now Zimbabwe and Zambia – in the 1890s, and was very invested in attempts to connect British colonies in Africa by rail.) From the beginning of their history, South African diamonds were inexorably intertwined with the history of empire, power, colonialism and capitalism in South Africa. And early experimentalists like Hannay would have been aware of the market for diamonds as their research overlapped so much with the influx of diamonds from colonies like South Africa.

In relatively short order, De Beers was everywhere in the diamond business. In London, it was known as the Diamond Tradition Company; in continental Europe it was the CSO (Central Selling Organization, which was an arm of the Diamond Trading Company); a few decades later, in Israel, it was known as 'The Syndicate'. At the height of its power, in the mid-twentieth century, De Beers either directly owned or controlled all the diamond mines in southern Africa (disguising its South African origins with subsidiary names like Diamond Development Corporation and Mining Services, Inc.), and owned diamond-trading companies in Britain, Portugal, Israel, Belgium, the Netherlands and Switzerland. In addition to the mines, De Beers controlled what gems went to what dealers. When Marilyn Monroe sang 'Diamonds Are a Girl's Best Friend' in the 1953 diamond-crazed film *Gentlemen Prefer Blondes,* De Beers had a complete monopoly on the diamond trade, controlling all mines and all sales of diamonds, and also had the financial and political resources to pre-emptively buy out any new diamond discoveries in just about any part of the world. There were no negotiations.

But, De Beers needed people to want to buy diamonds – and up until the nineteenth century only the very wealthy and aristocratic bought and wore them. 'The diamond invention is far more than a monopoly for fixing diamond prices,' journalist Edward Epstein argued in his 1982 investigative expose for *The Atlantic,* 'it is a mechanism for converting tiny crystals of carbon into universally recognized tokens of wealth, power, and romance.' De Beers hit on the idea of marketing diamonds as symbols of courtship and married life of the aspiring emerging post-Second World War middle class by promising customers that 'A Diamond Is Forever'. The slogan rolled out in 1947.

In the space of a mere two decades, De Beers convinced diamond-buying markets across the world that diamond engagement rings were the tangible symbol of affluence and success, disrupting hundreds of years of local cultural engagement customs. Until 1959, for example, diamonds were not permitted to be imported into post-war Japan. When De Beers began its diamond engagement ring campaign in 1968, less than 5 per cent of Japanese women who were getting married were given a diamond engagement ring – by 1972 that number had risen to 27 per cent, and by 1981 some 60 per cent of Japanese engaged women wore diamond rings. It took De Beers a mere 13 years to completely change the material culture of engagement in Japan. To ensure that people would not want to resell diamonds – to keep De Beers in the selling seat of the diamond market – De Beers found incredibly clever ways to imbue the gems with sentimental meaning. And year and year, decade after decade, the price of diamonds increased irrespective of the surrounding economic conditions.

In 1950, however, a threat to the De Beers diamond empire emerged. Scientists at De Beers knew that engineers and scientists in the United States (and the Soviet Union) were working on lab-grown, industrial-grade diamonds in labs like those of General Electric, and concluded that gem-grade diamonds would be next. The research lab took its concerns to Sir Ernest Oppenheimer (then the president of De Beers) to convince him that De Beers needed to seriously pursue the question of synthesising diamonds, as any outside patent would interrupt the company's monopoly. Oppenheimer dismissed the scientists and declined to fund the research, blithely declaring, 'Only God can make a diamond.'

Four years later, God apparently had no problem with General Electric cracking diamond-making technology. After General Electric's February 1955 press release, shares of De Beers stock plummeted. 'The diamond's price is strictly controlled by a European cartel,' *The Evening Independent* of Massillon, Ohio, reported in 18 February 1955. 'With General Electric now able to produce synthetic diamonds for industrial use – even if at a much higher price so far from that of a natural diamond – the time could come when the cartel's monopoly was challenged.' Investors were alarmed that nascent though the market was, it was only a matter of time before laboratory-grown diamonds outstripped the market for natural ones. Even more concerning was the possibility that these laboratory diamonds were outside De Beers's carefully controlled diamond ecosystem. In March 1955, Oppenheimer reversed his position and ordered De Beers's research lab to begin a programme to create diamonds.

The De Beers team began developing the high-pressure presses and catalysts necessary to produce diamonds, under the direction of Dutch-born physicist Dr J. H. Custers. But General Electric had a huge head start on De Beers, as the press and catalyst technologies took years to fine-tune. (De Beers's scientists reported that their lab was often 'rocked by explosions' and the walls were 'covered with smouldering carbon'.) De Beers began negotiations with ASEA in Sweden to buy laboratory equipment, and the US government came down with a tight embargo on publishing any details from General Electric's experiments, in order to freeze out any diamond-making research advances in the Soviet Union during the Cold War. It would be years before General Electric was permitted to even take out patents. By then, Tracy Hall had filed his own patents, having 'reinvented' the pressure belt to ensure that his company, MegaDiamond, wasn't in violation of General Electric's holdings.

After a period of trial and error, De Beers's scientists hit on the same conical press design that General Electric had used. Three years later, in 1958, the lab synthesised its first diamond from graphite. Custers declared 'Eureka' in his laboratory notebook.

Yet the results of the 1958 experiment were unrepeatable, a frustration that General Electric would have been well familiar with. It wasn't until 8 September 1959 that De Beers hit its stride with a method that resulted in success 60 per cent of the time – a few more tweaks, De Beers's logic went, and it would begin to look towards the possibility of taking out a world patent. At this point, General Electric's industrial diamonds were keeping pace with the price of De Beers's natural industrial-grade diamonds at $3 per carat, and the company knew that it had to act fast. General Electric pressured (as it were) the Eisenhower administration to lift the embargo, to enable it to file the patent for laboratory-grown diamonds before De Beers.

In mid-September 1959, General Electric filed its patents, cornering diamond manufacture before the South African team. After an extremely messy patent-rights trial in the early 1960s that lasted six years and consumed millions of dollars – where De Beers claimed that General Electric was in patent violation – the South African court ruled in General Electric's favour, and De Beers signed a licensing agreement to make laboratory-grown diamonds using General Electric's process and technology.

By the mid-1960s, lab-grown diamonds were pouring out of South Africa, the United States and the Soviet Union, and by 1970, more than half the diamonds produced in the world came from laboratories. Unlike the market for gem-grade diamonds – which De Beers tightly controlled and which continued to rise – the price for industrial-grade diamonds dropped sharply, buoyed up only because the world's consumption of industrial diamonds had actually quadrupled between 1955 and 1970 due to new uses for diamond abrasives. 'The decision a decade ago by De Beers to go into synthetic diamond production was obviously a highly emotional matter, akin to mounting an attack on one's own children,' the *Detroit Free Press* reported decades earlier, on 27 October 1969. 'But it had to come and De Beers decided it was better to be on the inside than the outside.'

When General Electric announced its one-carat, gem-grade diamond mark in 1970, De Beers reacted calmly, despite the fact that it had decided against trying to synthesise gem-grade diamonds, a decision it now regretted. It would seem that it was impossible to differentiate cut and polished General Electric diamonds from natural ones, even using a jeweller's loupe, crushing De Beers's carefully curated cultural invention of a diamond's significance. The only discernible difference, in fact, between General Electric diamonds and natural ones was that lab-grown diamonds tended to phosphoresce under an ultraviolet lamp. General Electric opted not to seriously pursue manufacturing gem-grade diamonds, citing concerns that if it did so the entire diamond market would collapse. In a 1970s' exposé about the diamond industry, a senior General Electric executive went on record with, 'We would be destroyed by the success of our invention. The more diamonds that we made, the cheaper they would become. Then the mystique would be gone, and the price would drop to next to nothing.'

By 1996, it was clear that between CVD technology and the increased production of jewellery-quality diamonds, synthetic diamonds were here to stay in the luxury market. To carefully maintain the divide between natural and synthetic diamonds, De Beers developed technologies that could differentiate lab diamonds from natural ones. Two of these – the DiamondSure and the DiamondView – were able to detect the presence of an optical absorption line, found in the majority of natural diamonds but not in laboratory ones.

These technologies clearly established an ethos that, yes, laboratory-grown diamonds were 'real' diamonds, as both were the same mineral. However, it reinforced De Beers's narrative that it was important, necessary and prudent to divide diamonds between natural and non-natural, and that De Beers's technology provided a way of maintaining that divide under the auspices of technology. If a diamond was to be forever, it simply wouldn't do to have science make it appear at will in a laboratory.

★ ★ ★

Diamonds have changed a great deal since the 1950s. For one thing, De Beers no longer holds a monopoly on the diamond market, natural or otherwise. After decades of price fixing and issues of legal trust, De Beers now only sells approximately 35 per cent of the world's diamonds. (In 2004, De Beers pleaded guilty to the 1994 US-brought charges of collusion with General Electric to fix the price of industrial diamonds, and paid a US $10 million fine. General Electric was acquitted of all charges.) But more than legal issues, De Beers experienced pushback from its customers. Having spent the majority of the twentieth century cajoling, manipulating and persuading customers that a diamond engagement ring was a cultural necessity of middle-class success, customers pushed back.

In 1999, a campaign led by Global Witness highlighted the role that diamonds play in international conflicts, and in March 2000 the famous 'Fowler Report' detailed the extent to which European and African governments and financial companies violated UN protocols, leading to an undeniable link between the illicit diamond trade and armed conflict in developing countries.

Known as conflict diamonds, war diamonds or red diamonds, these diamonds are mined in war zones and their proceeds are used to fund a plethora of unsavoury activities, such as insurgency, arms dealing and terrorism. They are extracted using underpaid labour in unsustainable mining practices in countries like the Democratic Republic of the Congo, Angola, Sierra Leone and the Ivory Coast. (Although illegal and unethical diamond mining practices are not limited to Africa.) De Beers stopped buying diamonds from conflict-zone areas in 1999, and by 2000 guaranteed that all of its diamonds were conflict free. In 2000, the Kimberly Process Certification was established to prevent conflict diamonds from entering the raw diamond market, although the efficacy of the Kimberly Process has been challenged.

The cost of conflict diamonds catapulted to public consciousness with the release of the film *Blood Diamond*, starring Leonardo DiCaprio, in 2006, along with a number of books, investigative journalism and high-profile publicity. Was it ethical, many consumers began to ask themselves, to own a diamond if it meant contributing to the trade in conflict diamonds? Laboratory-grown diamonds offer consumers an ethical alternative. Indeed, in 2015, DiCaprio (as well as other backers) invested in Diamond Foundry, a start-up company in Santa Clara, California, that grows gem-grade diamonds, publicising its jewellery as an ethical alternative to natural diamonds, conflict free or not. 'I'm proud to invest in Diamond Foundry, Inc. – cultivating real diamonds in America without the human and environmental toll of mining,' DiCaprio states on the Foundry's website.

De Beers countered with a campaign in 2016 that proclaimed 'Real is Rare', targeting millennial buyers – a less than subtle counterpunch, implying that laboratory diamonds were neither, since they could simply be conjured out of a lab. A friend of mine who got engaged while I was writing this chapter opted to propose with a laboratory-grown diamond. I asked him why he chose a lab diamond over a natural one. 'Laboratory diamonds have significantly lower environmental and social impact, which my girlfriend finds important. I decided that if I am to buy a diamond, then I'm not going to buy one that comes with an artificially high price,' he explained decisively. 'Besides, as a physicist, I liked the idea that the lab diamond was a better carbon crystal lattice.'

Customers and start-ups are finding even more unique ways to incorporate laboratory diamonds into everyday life. The same versatility of the carbon that allowed General Electric's Superpressure group to pull the crunchy-peanut-butter-to-diamond stunt allows twenty-first-century companies the opportunity to offer 'memorial gems', where cremated ashes of a loved one are trans-formed into a diamond. Companies like the start-up Eterneva in Austin, Texas, promise that a memorial diamond will be, 'an heirloom to be passed down through the generations'. When I interviewed the spokesperson for Eterneva, it was clear that the company sees itself as part of people's grieving process, and as a positive alternative to celebrate and cherish the life of a loved one. Memorial diamonds, much like non-natural diamond engagement rings, offer consumers an option of choosing how the cultural mores of diamonds can be rewritten, even after millennia.

Natural diamonds undergo a cultural transformation from mineral to gem – a narrative that imbues them with value and meaning. The story of diamonds in the twenty-first century is a story of capitalism – but unlike the capitalism of twentieth-century De Beers, contemporary consumers demand diamonds with different cultural values and ethics. It would seem that those ethics are coalescing more and more around non-natural diamonds – so much so, in fact, that in May 2018, De Beers announced that it was launching 'Lightbox', a line of lab-grown diamonds targeting the Sweet Sixteen necklace market, offering the less than natural diamond as a starter diamond for the 'real' one later on. Headlines the week of the launch proclaimed, 'A diamond is forever – and forever now costs $200 from De Beers.' Laboratory-grown diamonds have their own parallel process of social and technological transformations.

When General Electric demonstrated that it could create diamonds – and when its results were independently verified – the implications of laboratory-manufactured diamonds had concrete, tangible effects for the diamond industry. In the ensuing decades, synthetic diamonds moved from a hypothetical question of science and engineering into the tangible, with concrete implications for contemporary diamond markets. Today, diamond sellers are working to translate the material reality of lab diamonds into the appropriate metaphor and cultural mores for twenty-first-century consumers.

The twentieth century made laboratory-grown diamonds a reality. It's up to the twenty-first century to make them as authentic as natural ones – to make them the Real Thing.

CHAPTER FOUR

A Fake of a Different Flavour

In the mid-nineteenth century, very few Americans had ever eaten an actual, physical banana. The fruit was introduced in 1876 during the Philadelphia Centennial, a six-month-long celebration of the 100-year anniversary of the signing of the Declaration of Independence. Held in Philadelphia's extensive Fairmont Park, the Centennial attracted something like 10 million visitors, featured exhibits from 37 countries, and showed off a plethora of novel inventions and technologies. The event was also an opportunity for visitors to try new-fangled, exotic foods and flavours like Heinz tomato ketchup and Hires root beer. And, as history would have it, bananas.

One of the exhibits in the Centennial's Floral Hall showcased a cluster of green, leafy banana trees in a wooden crate from Central America where people could buy a banana for a dime. Although most visitors would have been eating the fruit for the first time, plenty of them would already have known what a banana tasted like – artificial banana flavour had been readily available to consumers for more than a decade.

Before the Centennial, bananas were a luxury food found only in affluent households. However, after the event, the public's demand for bananas skyrocketed in mere decades; bananas quickly became the first fruit to be commercially available year-round. By 1929, bananas comprised 50 per cent of United States imports from Central America. (In 1917, the Fruit Dispatch Company – an arm of the mega-conglomerate United Fruit Company – published a pamphlet for consumers called 'The Food Value of a Banana', which offered pragmatic and nutritional reasons to eat the fruit. 'Put up and sealed by nature in a germ-proof package.' 'Always in season.' 'Available everywhere.' 'The poor man's food.' 'Endorsed by physicians.') However, the increasing popularity of the fruit did little to diminish the appeal (as it were) of artificial banana flavour.

Thanks to the growing popularity of artificial flavours in the mid-nineteenth century, Americans had come to expect a certain sticky-sweet taste from any banana-flavoured food. Chemists, pharmacists and early flavour tinkerers found that the chemical compound isoamyl acetate could elicit a broadly 'fruity' flavour when it was added to foodstuffs like sweets and puddings. Today, isoamyl acetate is in fact recognised as one of the main banana esters – one of the primary components – that characterises banana flavour. The first formulas for synthetic banana flavouring date to the 1860s, and advertisements for 'fruit essences' that could be used to create flavours – including banana – go back to as early as the 1850s.

By the mid-nineteenth century, American chemical suppliers began to market the isoamyl acetate compound as 'banana flavour' to pharmacists and confectioners who were looking to expand the scope of manufactured flavours; consequently, the sensory association between isoamyl acetate and bananas quickly became entrenched in American consumers. So, when visitors to the Philadelphia Centennial plunked down 10 cents for a banana, they found that the banana tasted like, well, a banana. The fruit, with its high concentration of isoamyl acetate, confirmed to consumers that the artificial banana flavour tasted like the Real Thing. Decades later, when flavourists were able to match specific chemical compounds to specific flavours of foodstuffs, isoamyl acetate was one of the first chemical compounds used in artificial flavour that was confirmed to exist in actual fruit as well.

Today, it's hard to imagine that the sickly-sticky-sweet-synthetic banana flavour of Laffy Taffy sweets has anything to do with real bananas because they taste so different from the fruit. However, today's particular taste discrepancy has more to do with the bananas we consume in the twenty-first century and less to do with the assumption that synthetic banana flavour is a poor approximation of the real fruit.

When bananas became significant imports in the early twentieth century, the Gros Michel banana quickly came to dominate the banana industry. Purportedly from the Caribbean island of Martinique (north-west of Barbados), the Gros Michel was an ideal banana for shipping to international markets, because Gros Michels had a long ripening period, and a thick, leathery peel that ensured that the fruit wouldn't easily bruise; the individual bananas in a Gros Michel bunch also grew very close together, giving exporters the most cost-effective banana to shipping space ratio. Consequently, when someone was talking about a banana in the early twentieth century, they were by default talking about a Gros Michel banana.

However, only one variety of banana at a time is cultivated for the global market, despite there being more than 1,000 varieties of banana around the world. This strategy is efficient, yes, but it's extremely risky. With everything riding on one banana type, that one banana type is very susceptible to disease. Midway through the twentieth century, the Gros Michel banana effectively went extinct, when growers became unable to combat the fungal plagues that wiped out the variety. The Cavendish banana, with its iconic bright yellow skin, replaced the Gros Michel as the mono-cultivated banana crop, and today Cavendishes are the most common bananas in the global banana market. However, due to the way that Cavendishes are cultivated – they're seedless and therefore sterile – they are essentially clones and highly susceptible to disease, since they have no genetic diversity to fall back on. Today, the Cavendish banana is under threat from disease, and scientists consider that the question of its extinction is not 'if' but 'when'.

But – and this is significant to the story of artificial banana flavour – not all bananas taste the same. The Gros Michel and the Cavendish are in fact very different-tasting varieties of banana. Gros Michels contain more isoamyl acetate compound than Cavendishes, making Gros Michels taste more 'banana-y'. Consequently, tasting synthetic banana flavour, codified when the Gros Michel was the commercial banana *de jour*, is like tasting a relict banana from the last century. Even though contemporary flavour scientists would be able to more accurately copy the flavour of today's bananas, consumers expect artificial banana flavour to be what it's always been.

When the Cavendish banana goes extinct and is replaced by some other banana variety, the cultural perception of banana flavour will evolve yet again. 'The fake fruit flavors that we encounter in some of the most common frozen treats, and sugary candies, and cheap sodas, kind of take us back to the early days of synthetic flavor,' flavour historian Nadia Berenstein explains in an online interview with NPR's *Science Friday* programme. 'They give us a kind of glimpse into the flavor worlds of the past.'

This consumer-based commitment to what Berenstein calls 'heirloom synthetic flavors' offers a reminder that flavour – especially synthetic flavour – is complex and its history even more so. Although 'natural' and 'artificial' carry specific meanings in the flavour and regulatory industries, finding where natural flavour leaves off and non-natural flavour picks up requires a willingness to believe that the flavour of something can change and evolve.

★ ★ ★

The story of flavour actually begins with an understanding of taste as well as how people perceive the phenomenon of flavour. How something tastes is part biochemistry, part neurophysiology, and a very large part history and culture.

On a biological level, taste is one of the traditional five human senses, and when taste is combined with smell and cranial nerve stimulation, we experience flavour. The tongue is covered with thousands of bumps called papillae – invisible to the naked eye – and each papilla is filled with hundreds of taste buds. Other taste buds are located on the roof and sides of the mouth, as well as in the throat, and when we put food in our mouths it reacts chemically with the taste receptor cells in our tongues. People with a much higher than average number of papillae on their tongues are known in the world of flavour science as 'supertasters'; the extra taste receptors make them much more sensitive to nuances of flavours in different foods than non-tasters, those with average taste buds. Taste status – like odour sensitivity – is heritable. (In fact, in the 1930s some scientists advocated using a taste test as a paternity test, although this didn't really catch on.) In addition to the chemical reactions of taste in papillae, taste also depends on our smell as well as stimulation of the trigeminal nerve (the largest cranial nerve), which in turn processes how food's texture and temperature register as a sensation within the mouth.

More than 2,300 years ago, the philosopher Aristotle postulated that the two most basic taste sensations were sweet and bitter. This short list of taste categories has expanded in the subsequent millennia to include sweet and sour, bitter and savoury, salty and – although this is still under debate – fat. Recent flavour researchers would also include 'umami', which means 'delicious taste', first described by the Japanese researcher Kikunae Ikeda at the beginning of the twentieth century. (Umami is generally taken to mean savoury, brothy or meaty.) Other types of taste sensation include coolness, numbness, metallicness, astringency, pungency, calcium, heartiness and starchiness. Taste research has even shown that temperature plays a key role in determining the taste of a particular flavour.

For decades, flavour experts thought that people tasted sweet at the front of the tongue, bitter at the back and sour at the side, in a pattern called a 'tongue map'. However, recent research has shown that every taste bud is responsive to all five of the basic tastes. The issues – misconceptions – with the tongue map can trace their history back to a mistranslation from German that first appeared in Edwin Boring's 1942 popular psychology textbook. Biochemical research has identified that bitter, sweet and savoury depend on G-protein-coupled receptors – however, the receptor for sour remains unknown and undefined.

Although the same biochemical processes are at work in every person's body, there are several genetically determined differences in what a person tastes and how they perceive certain chemical compounds. For example, for some people cilantro (or coriander) tastes fragrant or citrus-like, while for others it tastes soapy. Some people are able to detect bitterness in foods on a more sensitive scale than others. Something like one in every two people is unable to smell androsterone (a particularly pungent steroid derived from testosterone) and estimates put 35 per cent of the world's population registering androsterone as an incredibly unpleasant smell similar 'boar-taint' flavour. Roughly 1 per cent of the world's population is unable to smell vanilla. In short, for all of the biochemistry that goes into how taste works in the human body, there is an awful lot of genetic variation in just how those taste receptors fire.

Speculation about the role that flavour – taste – has played in human evolutionary history runs rampant. 'Human taste abilities have been shaped, in large part, by the ecological niches our evolutionary ancestors occupied and by the nutrients they sought,' geneticist Paul Breslin offers in an article about the evolution of food and human taste in *Current Biology*. Some researchers have suggested that parsing flavour was crucial to determine foods that were safe to eat and those that weren't. Plants with poisonous toxins would taste bitter, for example. Rotting meat would have tasted so noxious to our evolutionary ancestors, anthropologist Richard Wrangham argues, that the taste would have served as a deterrent to eating meat before it was properly cooked.

Although poor preservation often stands as a yardstick of meagre culinary hygiene, rotting meat doesn't necessarily indicate a failure of food or taste. Contemporary anthropologists have found numerous instances where fermented and rotted animal foods, like the stomach contents of certain ungulates and birds, are – or at least were until very recent history – dietary staples vital to the success of foraging groups in extreme northern climates. Consequently, archaeologists have begun to speculate that putrid meat and fish may have played a significant role in the culinary history of Europe's Middle Palaeolithic for both Neanderthals and *Homo sapiens*, between 45,000 and 300,000 years ago. (That, dear reader, is a genuinely authentic paleo diet.) Regardless of anything else, it's clear that flavour, cuisine and diet occupy an ever-changing balance in what humans eat and why they eat it.

What is particularly unique, however, to mammalian evolution – and an evolutionary harbinger to flavour – is the mammalian nose, an organ that can detect volatile odours. Although smell does not seem to be as significant in primate evolutionary history as vision, olfaction is the oldest evolutionary part of the sensation of taste. Even the eighteenth-century naturalist Carl Linnaeus, famous for devising a classification system for the world's plants and animals, invented a taxonomy for odours, arguing that all odours could be categorised as fragrant, spicy, musky, garlicky, goaty, repulsive or nauseating.

Today, flavourists who focus on olfaction suggest that there are four dimensions of smell; namely, fragrant, acid, burnt and goaty. 'Smell is an involuntary ubiquitous sensation. Waking or sleeping, eyes shut or open, we cannot help but smell with inspiration,' molecular evolutionary anthropologist Kara Hoover says in an overview of the evolutionary significance of olfaction in the 2010 *Yearbook of Physical Anthropology*. 'The other senses can be stopped manually (closing the eyes or plugging the ears), but we cannot stop breathing; even mouth-breathing will impart a weak sense of smell.' While it's incredibly easy to build up plausible stories about how 'our flavour sense may have played a large role in making humans into the species we are,' the significance of flavour to human evolution – and specifically, of flavour preferences, trends and histories – is a very recent branch of evolutionary and flavour research.

Fast forward millions of years in mammalian evolution and we come to researchers studying the neuroscience of flavour, working out how chemical compounds translate into flavour experiences. So taste is about biology, of course. Biochemistry, great. But how the brain – or mind, if you prefer – interprets flavour input is the final component of flavour perception. 'In other words, it is as much a matter of what is in the *mind* of the person doing the tasting as what is in the *mouth* or on the plate,' flavour and sensory scientist Charles Spence points out. In *Gastrophysics: The New Science of Eating*, Spence relates a story about star chef Heston Blumenthal who, in the late 1990s, created a crab-flavoured ice cream to accompany his popular crab risotto main dish. The savoury ice cream really worked for Blumenthal as a quirky but savvy complement to the rest of the meal. But a crustacean-flavoured ice cream dessert did not, shall we say, go over well with diners.

Part of the problem, it turns out, was that the ice cream was pinkish-red in colour – a hue that people have come to expect to be associated with sweet, fruity flavours. Blumenthal's diners presumably thought that they were going to taste something sweet, and what they bit into was the complete flavour opposite. 'In other words,' Spence wryly noted, 'they were expecting strawberry and got frozen crab bisque instead!' In a series of experiments, psychologist Martin Yeomans has shown that it is possible to influence a person's perception of liking a frozen pink dessert by simply telling them that it is savoury to begin with – or giving the food a different name. The crab-flavoured ice cream tasted vile, Yeomans' research showed, in large part because it was so completely different from what a person expected the flavour to taste like. If people's expectations of flavour weren't disappointed, then the flavour tasted 'better'.

Likewise, flavour is affected by a food's provenance. It turns out that when people like the story of their food's origin, they are more likely to think that it tastes better. One recent study that Spence cites offered identical samples of meat (like beef jerky or ham) to study participants, telling them that the meat was either factory farmed or free range. Those who were told that their meat was factory farmed 'rated it as tasting less pleasant, saltier and greasier'; they ate less of it, and said that they would be less willing to pay high prices for it. Crucially, this pattern held up throughout three separate studies, including one that found the inverse effect – people who believed that they were eating free-range, organic meats consistently thought that the meat tasted better. In blind taste tests, however, consumers cannot – for the most part – tell the difference. To that end, research has found that oysters taste better with the sound of the seashore playing in the background. 'So what this means in practice is that if you shell out for some organic, free-range, hand-fed food, you should be sure to let your guests know its provenance if you want them to be able to taste the difference,' Spence quipped.

So, biology, great. Biochemistry, of course. Neuroscience, indisputably. But taste – and flavour – also depends very much on the culture consuming it. Although the perception and expectation of flavour happens on an individual level, culture writ large facilitates our expectations of what we think something 'ought' to taste like. A classic example would be temperature-mediated taste – in the United States, soft drinks 'ought' to taste cold, regardless of season or anything else. And since the culture surrounding the production and consumption of foodstuffs changes, it's not a reach to conclude that flavour changes over time as well.

Since taste changes over time in the ever-evolving nature of a food's flavour, there is often a gap between how food tastes and how people think it ought to taste – particularly for contemporary audiences. Most flavourists and food historians describe this phenomenon as gastronomic nostalgia – a wistfulness for the storied provenance of the food. This is how people long for something to taste authentic, even if they've never had the 'real' or the 'original' flavour.

When people lament that food tastes differently than it did in previous generations, 'people usually mean that fruits, vegetables, bread, beer and meats are not what they once were – not as tasty, not as authentically what they are supposed to be,' historian of science Steven Shapin observed in a formal lecture called 'Changing Tastes', given at Uppsala University in Sweden in 2011. Shapin offered several historical reasons for why flavours have changed over time. New varieties of foodstuff, for example, might not be bred for exactly the same tastes as ones from earlier centuries, or cultures may have lost the art of preparing certain foods in ways that enhance particular flavours.

Some foods – like certain varieties of apples and pork – have effectively disappeared from contemporary food supplies. Other foods have changed what we think they ought to taste like. But these factors don't stop people mourning their loss – or more appropriately mourning what they think has been lost, because so much of flavour and taste have to do with nostalgia more than anything else. 'Gustatory nostalgia is very much on the late modern menu,' Shapin points out.

Culture gives us the surrounding ethos of taste and how it is created, cultivated and considered – culture is how taste is transmitted and how flavours are formalised into everyday expectations.

★ ★ ★

Flavour has always been guided – engineered – to carry specific messages about taste and status to consumers. For Europe of the late Middle Ages and early modern period – the fourteenth, fifteenth and sixteenth centuries – spices were the most significant factor in shaping food tastes. Over these three centuries, spices were valued commodities in European markets and were the impetus for Europe's maritime expansion. (Say it with me Frank Herbert and *Dune* fans, 'The spice must flow.') The twentieth-century French philosopher Pierre Bourdieu referred to the phenomena as *habitas* – the physical embodiment of cultural capital. Thanks to spices, flavour curried a certain cultural cachet.

But how did these seasoning flavours come to shape European cuisines and economies so decisively? One older, popular opinion holds that spices were used to preserve meat as well to hide the taste of rotting foodstuffs during the period when Europe was struggling to eke its way out of its medieval past. Other food historians argue that Europe's interest in spices lay in the demarcation of social distinctions – the idea that aristocrats higher up on society's socioeconomic ladder would be able to buy and ply spices in a way that mere peasants couldn't. Still other experts suggest that the West learned about spices from Arab cultures during the Crusades, and when crusaders brought the spices back to Europe and, subsequently, to European cooking. Regardless of precisely why, spices became a significant component in the shaping the mid-millennia tastes of European food, proof that flavour could be both altered and directed.

In addition to controlling flavour, spices were used for medical and therapeutic purposes, giving certain foods medicinal distinction as well as being something to simply consume. (For example, according to *Le Thresor de Santé*, a 1607 French compilation of notes about health and medicine, cloves were purportedly good 'for the eyes, liver, hearth, and stomach', and pepper 'facilitates urination … cures chills from intermittent fevers and heals snake bites'.) Every spice in a medieval kitchen was originally imported for its medicinal qualities, and only later was it employed as a seasoning. Because spices were medicinal, the use of spices – and food they spiced – was also imbued with moralisms about health. Both health and flavour continue to shape how people think their food ought to taste, and also how their food obtains that flavour in the first place; this dualism of health and wellness helps enforce the idea that taste can have a higher moral purpose.

But what if we could shape flavour outside of such traditional methods like agriculture or spices? What if it was possible to synthesise – in a laboratory – flavours that had only ever been found in nature?

Beginning in the nineteenth century, flavour moved from being 'guided' to being outright manipulated, copied, replicated and invented. Flavour engineering in the nineteenth century was a way of mimicking nature, of invoking familiar flavours and tastes, through chemical manipulation of what went into food.

By the turn of the twentieth century, flavour had become something that was copied or replicated from nature, domesticated nature though it might be. Flavour was distilled down to its chemical components, which were then codified into standard flavours that were, in turn, put into foods. With the emergence of foodstuffs that were designed to be produced and consumed on a massive population level – like tinned vegetables – the question of how to take the flavour from one food and effectively put it into another began to underscore flavour science. In the second half of the twentieth century, it wasn't enough to simply replicate nature's flavours – flavourists wanted to engineer flavours that were even better than their natural counterparts or to simply design new ones.

Synthetic flavours – more commonly called artificial flavours – were invented in the late nineteenth century, and were heavily tied to the perfume industry and the chemical research of artificial scents. 'Once an odor is experienced along with a flavor, the two become associated,' *Scientific American* explains of the two closely related sensations, 'thus smell influences taste and taste influences smell'. If you wanted to create artificial flavours, it would make sense to piggyback off research to do with artificial smells.

These new synthetic flavours were not invented to satisfy a pre-existing need, or only as inexpensive substitutes for the 'real thing'. As a result of advances in chemistry, the flavour of bourgeoisie foodstuffs could be distilled, created and infused into processed foods that were produced for long shelf lives and at a scale that could be widely available to the burgeoning middle class of

America. The story of synthetic flavours is one of industrial science, chemistry and mass-consumer economy – the stories of changing tastes.

Early flavour designers depended on the sensory research of perfume chemists and manufacturers of the late nineteenth century. While these scientists were interested in creating artificial smells that mimicked odours in the real world, their counterparts in the flavour world were starting to explore how to apply that chemical knowledge to transforming, enhancing and adding flavours to different foodstuffs. The flavour and perfume industries had long used 'essential oils' – made up of parts of a plant that could be distilled, extracted, then infused into something else, like food or fragrance. Archaeological evidence suggests that ancient Persians distilled essential oils as early as 3000 BC, and 2,000 years later Arab civilisations rediscovered and refined the process. The point is that the idea of distilling oil to its most essential building block and infusing it into something else is nothing new. Some of the first examples of an artificially created flavour came from using 'ethers', sometimes known in the industry as 'fruit oils', and both of these grew out of the tradition of distilling and utilising essential oils.

Combining chemistry with confectionary making was not an uncommon nineteenth-century practice. In 1855, for example, the confectioner Mr Samuel Simes of Philadelphia published an ad for artificially fruit-flavoured sweets made in his store – a four-storey building that took up the north-west corner lot on Chestnut and 12th Streets in Philadelphia. His retail drugstore was also a chemical manufacturing business – the flavours as well as the sweets were produced 'in house'. The ad boasted that the fruit essences he manufactured 'expressly for confectioners' gave sweets 'the rich and luscious flavours of the different fruits more decidedly than the fruits themselves'.

The synthetic flavours that Simes offered included Pineapple, Strawberry, Raspberry and Jargonelle Pear, as well as Vanilla, Orange and Blackberry. The story of Samuel Simes's synthetic flavours was in keeping with pharmacy flavour-making practices of the 1850s, as supply houses and chemical-supply catalogues began listing 'compound ethers' among their products – sometimes with a descriptive, easily understood 'apple oil' alongside the standard chemical name, like amyl valerate. The fruit smell associated with each flavour came to quickly define each compound scientifically as well as commercially.

Fast-forward six decades from Mr Simes's soda shoppe and we find ourselves with the invention of fake grape flavour. Nadia Berenstein tracked down the origin of today's fake grape flavour to 1910 or 1911, when a maker of flavouring extracts was riding a streetcar in Indianapolis and the scent of Concord grapes wafted from the perfume of a woman sitting next to him. According to Berenstein's archival sources, the flavour extract expert knew immediately that the smell of grape perfume could translate into grape flavour. 'This was important to him because he was a maker of flavoring extracts for bottled sodas and for soda fountains,' Berenstein explained in her *Science Friday* radio interview. 'So, he went through chemical catalogs, and he found the chemical in question, methyl anthranilate.'

Methyl anthranilate was marketed in Germany and Austria as the scent of orange blossoms, not grapes, in a bit of quirky cultural history that brings us back to the question of culture and perceptions of smell and taste. What was marketed as orange blossom scent in one cultural context could smell like grapes in another – and not just any grapes, but Concord grapes. 'I think the reason is that methyl anthranilate is a flavor compound in American grapes, in *Vitis labrusca* grapes, which include Concord grapes,' Berenstein offers. 'And it's not found in European grapes, in

Vitis vinifera grapes. So, the manufacturers of this synthetic compound in Germany and Europe wouldn't have noticed the resemblance with grapes because they were eating a different kind of grape.' Today, our fake grape flavour – codified forever in Americans' Jolly Ranchers and children's cough syrup – is this same chemical compound from the early twentieth century. Thus, the very real Concord grapes taste like fake grape and vice versa, reinforcing how taste and flavour are artefacts of past tastes.

Until the early twentieth century, there were a few, limited chemical compounds like amyl acetate, amyl valerate and butyric ether – pear oil, apple oil and pineapple oil respectively – that could claim to be legitimate substitutes for the 'real' fruit flavour taste that they invoked in their consumers. These ethers were little more than essences of fruits and other foodstuffs, and pharmacy trade journals readily published recipes and how-to guides for pharmacists interested in making artificial flavours. For example, Austrian chemist Vincenz Kletzinsky released his Table of Formulas for 'Artificial Fruit Essences' to the German-speaking world in 1865, and in translation to American chemists in 1867 – the table was Kletzinsky's latest pure and applied chemical research of distilling, and then replicating flavours.

The earliest artificial flavour recipes were discovered almost haphazardly, but were systematically studied and copied once their efficacy was demonstrated. For decades – between the middle and late nineteenth century well into the turn of the twentieth – chemists had a formalised and codified way to ensure that the artificial flavours they were making could be replicated by others.

Handbooks of artificial flavour recipes were compiled and sold. Books like *Elixirs and Flavoring Extracts: Their History, Formulae, and Methods of Preparation*, published in 1892 by J. U. Lloyd (professor of chemistry in the Eclectic

Medical Institute and former professor of pharmacy in the Cincinnati College of Pharmacy) were the early flavour industry's codification of how to create certain flavours. *Elixirs and Flavoring Extracts*, for example, had recipes for 'elixir of beef' as well as 'elixir of blackberry'. Both recipes called for extracts of the real food – be it beef or blackberry – and then directed the elixir maker to add 'simple elixir' with other chemical compounds like carbonate of magnesium. Many of the created flavours, like Coca-Cola, were originally formulated to treat everything from dyspepsia to headaches, and contained a variety of essential oils that included lemon, lime, orange, cinnamon and nutmeg in addition to cocaine, thus harkening back to the mid-millennia mix of health and flavour serving dual purposes.

A shift in flavour coincided with the mass production of foods in the United States and Western Europe. Over the twentieth century, artificial flavour became more important in making mass-produced foods more natural tasting and more palatable to consumers. No one expected these early flavour engineers to completely replace natural flavours – in part because they lacked the flavour technology to do so. The process of diagnosing, describing and creating artificial flavours became more formal as well as much more technical. Flavour quickly became an industry that was capable of producing taste and expectations of taste on a mass scale. In order to create synthetic flavours, however, flavourists need new tools, new methods and new instruments, like the liquid gas chromatograph, to keep pace with the industrialisation of flavour and the mass production of food. All of these moved the story of synthetic flavour from beyond the elixirs of the nineteenth century into the chemical industry of the twentieth.

In 1949, flavourists S. E. Cairncross and L. B. Sjöström of the Food and Flavor Laboratories of Arthur D. Little, Inc.

introduced the concept of 'flavour profiles' to the tenth annual meeting of the Institute of Food Technologists. This audience of flavourists had been grappling with how to describe created flavours for more than two decades. The idea of a flavour profile offered a descriptive analysis of flavour in a common language. 'First introduced in the late 1940s, the flavor profile was both a technology of flavor measurement and a powerful tool for flavor design – one that claimed the unique ability to detect and predict the qualities that would make a flavor successful among consumers,' Nadia Berenstein described in her PhD dissertation. 'Produced by a specially selected, highly trained sensory evaluation panel, a flavor profile was understood to be an accurate, comprehensive record of a substance's subjective sensory qualities.'

Now *de rigueur* foodie vernacular, flavour profiles were significant, formal tools in the mid-twentieth century. The labs of Arthur D. Little, Inc. quickly caught on to the idea that a flavour profile was both a concept of flavour and a method of measuring it. The flavour profile might be qualitative, but at least it would be consistent. Cairncross and Sjöström thought that they ought to describe where a flavour sat along a sort of flavour spectrum. A 1957 worksheet for profiling a malt beverage included a section about aroma and flavour by mouth, as well as the aftertaste. Respondents were asked to rate the intensity and amplitude of the fruity (that is, apple) aroma against the bitter (metallic) flavour and the CO_2 tingle. Other worksheets included descriptors like 'eggy' or 'rubbery', 'cabbage-like' or 'skunky', to describe a flavour's naturally occurring odours due to sulphides and organic sulphur compounds. In a worksheet for beer, the flavourists had respondents grade the skunkiness of both the aroma and flavour as part of building the profile. To formalise a profile, one person in the panel of formally trained tasters, Cairncross and Sjöström suggested, could act as a moderator and recorder to create a consensus of a group's responses. Putting together the responses of all of those trained flavourists – a group of 6–10 – meant creating a generally agreed upon formalisation of what a food – a strawberry, a raspberry, beer, whatever – tasted like.

Just as one set of flavourists in the industry was honing a method for how to describe flavour, other flavourists were busy actually creating flavours to add to food. Even today, there's an oft-cited theory that the human olfactory repertoire can detect something like 10,000 odorants, but that figure was simply based on an estimate with faulty assumptions from a chemical engineer at Arthur D. Little, Inc. in 1954. It's hard, in fact, to overemphasise the longstanding impact of flavour research from Arthur D. Little, Inc. in the early days of flavour synthesis.

Although flavour profiles were an important step in formally codified specific flavours, the introduction of gas chromatography (GC) in the 1950s allowed the new class of professional flavourists – as well as flavour chemists and perfumers – an increased ability to characterise and manipulate flavours and odours. Gas chromatography is a method of separating the different parts of complex compounds (like flavour) into their individual, molecular components. Most significant to the story of flavour, however, was the fact that, unlike traditional chromatography, gas chromatography used gas as a carrier agent to create the compound's molecular readout. Being able to analyse substances using gas meant that compounds that existed only in a vapour phase – like odour – were now available for flavourists to work with.

In short, gas chromatography translated the qualitative parts of a flavour profile and codified those components into a recipe of molecules. These were called 'flavour notes' in flavourist circles. When flavourists reproduce a flavour, they are using chemicals known as 'flavoxates'. In the 1920s, there were something like 70 family-owned essential oil and aroma chemical-flavour companies in the United States, more than 50 in Lower Manhattan alone. By the 1970s, more than three-quarters of them were out of business, due in large part to the introduction of methods like flavour profiling and the industrialisation through the use of the gas chromatograph.

The implications of the use of gas chromatography for flavour research were immediate and far-reaching. Here was a way of creating flavour that could be chemically identical to what researchers had found in nature! 'Researchers understood gas chromatography as the hot new technique for unlocking nature's secrets, an understanding furthered by increasing access to gas chromatographs,' food studies specialist Christy Spackman summarises in her research. Here were all the essential chemical building blocks of taste, all set for flavourists to duplicate and manipulate them.

Parsing the complexity of notes found in flavours offered up some unexpected findings to mid-twentieth-century flavourists. There was so much more to a food's flavour than seemed to meet the taste buds. Early use of the gas chromatograph showed flavourists where they needed to add unexpected notes. Moreover, several artificial flavours require rather unappetising notes. Flavourists found that they needed to add 'sweaty notes', for example, which are essential to imitation rum and butterscotch. 'Faecal notes' offered a full-bodied edge to both cheese and nut flavours. Processed fruit flavours needed a 'burnt' undertone to best mimic the effects of cooked fruit, while artificial tinned tomato flavours must include the 'tinny' taste to meet audience expectations of what a tomato 'ought' to taste like.

It's easy to assume that once gas chromatography was combined with the technique of flavour profiling, creating particular flavours was both straightforward and routine.

Although gas chromatography might guide the creation of molecular compounds needed to recreate a particular flavour, it's not foolproof. It turns out that knowing what flavour notes make up the symphony of how a particular food tastes isn't enough to crank out a replica of its flavour.

Most flavourists tried to reproduce familiar flavours – or at least to inspire a taste of something that's familiar to consumers – and very few tried to create genuinely new flavours. 'In the 1960s, however, flavourists set out to recreate virtually the whole spectrum of food flavours, from fruits and vegetables to meats,' historians Constance Classen, David Howes and Anthony Synnott point out. 'They have not been completely successful: some flavours notably chocolate, coffee and bread, have eluded accurate simulation.'

Although the gas chromatographs gave researchers a readout of a flavour, it turns out that there are a lot of 'unnecessary' notes in the chemical make-up of flavours. Coffee, for instance, contains more than 800 different flavour compounds – only a small number of which are considered absolutely essential to coffee's characteristic flavour. Also, certain flavour notes might be missing, and proportions of compounds might be wrong – maybe the synthetic molecules flavourists are using contain impurities when they are synthesising the flavour, impurities perceptible to the nose but not the machine, for example. Flavours made by relying only on readouts from a gas chromatograph may functionally fail to perform as needed in foods – a flavour in nature may involve molecules that are highly unstable, prone to oxidise or degrade, or are super volatile. In other words, there's an awful lot of art to guide the science of flavour, just as there has been for millennia.

For most of the artificial flavour development of the 1950s–'60s, flavourists pushed the idea that it was possible not only to replicate flavour found in nature, but also to improve it. An American flavour company bulletin from the 1950s announced, 'We are proud to announce our new improved cherry flavor, of course it is still no match for Mother Nature's.' But what if a flavour could be designed from the start? What if you could make a strawberry flavour taste more like strawberries than any garden-grown strawberry could? Or what about a flavour that had never been tasted before? What if, instead of copying nature's flavours, scientists could improve them?

This was an important turning point in the story of artificial flavour. No longer satisfied with humbly mimicking nature, mid-twentieth century flavourists set themselves up squarely in opposition to the natural world. Emboldened by decades of success, flavourists in the 1960s began to experiment with actively creating new flavours and driving consumers' expectations about what food ought to taste like. Mother Nature might have given us strawberries, the flavourists' logic went, but by golly there's no reason why we can't make strawberry flavour taste even more strawberry-like.

As consumer expectations about flavour changed throughout the decades of the twentieth century, flavourists rushed to help shape how we think food ought to taste. It's as if the flavour industry was taking directions from Willy Wonka's other-worldly flavour making – nothing, it would seem, was too fantastical to try. If you could envision the flavour, you could make it.

Distilling out the components of flavour – to describe flavour in its chemical or molecular components – is one thing. Being able to pare down that chemical compound to its essential parts is another. It turns out that, 'Nature has a way of decoying its secrets with unnecessary ingredients,' as flavourist Charles Wiener suggested in the 1980s. It was his job, as was that of his contemporary flavourists, to parse what parts of foods yielded what sorts of flavours and how best to reproduce them. ('I think that is the best blueberry flavor that's ever been made,' Wiener told journalist Ellen Ruppel Shell when she toured his laboratory at International Flavors & Fragrances Inc. for a profile she wrote for *Smithsonian Magazine* in May 1986. 'And there's not a scrap of blueberry in it.') For consumers, it wasn't – isn't – enough to reproduce flavour as the exact chemical make-up that exists in natural foods. It must also match what consumers expect something to taste like. This required a new, specific, formal set of methods for creating and describing flavour.

This shift in consumer flavour expectations underscored that flavour is anything but static. Once a specific flavour was catalogued – say, 'D & O 5210 Strawberry' – then that flavour became the industry standard for what constituted a flavour. (Historically, in 1939, when flavour researchers were first working out the chemistry of strawberry flavour, researchers started by pressing out 445kg [980lb] of juice from physical strawberries to determine a strawberry's essential chemical components.) However, just because a flavour became codified through a formula, this didn't mean that flavourists wouldn't still tinker with it. It turns out that strawberries, like so many complex flavours, are in fact incredibly complicated to create because they don't have just one or two dominant notes in their flavour, and new flavour notes are still being discovered.

In 1992, prominent flavourist James Broderick pointed out that although the 1950s' D & O 5210 Strawberry was the industry standard, he and his colleagues were attempting to match a slightly different strawberry symphony. Creating a 'new' industry standard for strawberry meant introducing 'green' notes to the flavour compound. 'The demand for "natural" flavors and the availability of "natural" ingredients (such as ethyl butyrate, ethyl 2-methyl butyrate, diacetyl …) enable the flavorist to produce a "natural" strawberry with a minimum of juice or fruit extractives,' Broderick explained in his writings. He concluded by pointing out that aldehyde C16, once thought to be essential to the manufacture of strawberry flavour, was no longer considered necessary. So, according to Broderick, we take out the aldehyde C16, add some green notes and *voila!* we have a new – but still natural – strawberry flavour. A strawberry's flavour profile was thus then considered a mix of fruitiness, balsamic, rose-honey, greenness, rose, butter, straw and sour. Synthetic strawberry was strawberry to begin with, but now it was even more so.

In a 2009 *New Yorker* profile of the twenty-first-century flavour business, journalist Raffi Khatchadourian' pointed out that flavourists were still tinkering with strawberry flavour. 'There is a note that adds overripe to strawberries,' flavourist Michelle Hagan told Khatchadourian. She pointed out that a flavour note labelled '2-OCTEN-4-ONE' from her laboratory supplies was in fact a recently 'discovered' compound in physical strawberries and thus a significant, but previously unknown, component of strawberry taste.

The line between natural and artificial flavour – in terms of what that flavour tastes like – has become more and more fuzzy. Less than a decade ago, the molecule was considered artificial, but it was chased down in the natural world and its status changed. '2-OCTEN-4-ONE' is now natural – a powerful reminder that the boundary between artificial and natural is both fluid and arbitrary.

★ ★ ★

There are few foods that dance around the ever-evolving gradient of flavour engineering quite the way that Jelly Belly jelly beans do.

In 1866, Gustav Goelitz immigrated to the United States from Germany. Three years later, with the help of his two brothers, he started the Gustav Goelitz Candy Company in Belleville, Illinois – today, we know the company as Jelly Belly. The Gustav Goelitz Candy Company started out making 'mellowcreme' candies, and at the turn of the twentieth century was widely credited with the invention of candy corn. In the early 1960s, the company (then called the Herman Goelitz Company, after the fourth-generation descendant of Gustav) expanded its inventory to include gummies and jelly beans, crafting a small 'natural' flavoured jelly bean, and in 2001 it renamed itself outright as the Jelly Belly Candy Company. It's hard to imagine that when Goelitz started out in 1869, he could have conceived of a world where, 150 years later, that same company would have created more than a hundred flavours of jelly beans, and where Booger and Vomit flavoured jelly beans would not only sell – but would sell well.

Jelly Belly beans are not just any old jelly beans. The jelly beans got their start in 1965, when Goelitz started manufacturing the Mini Jelly Bean, a small sweet that had 'natural flavour' infused into the centre of each piece, setting these beans apart from those of other confectioners, who only flavoured the outer shells of jelly beans, and those outer shells were artificially flavoured at that. In 1976, a confectioner named David Klein began collaborating with Herman Rowland of Goelitz to create a jelly bean that would use a 'natural puree' to fill the jelly beans. Those first Jelly Belly flavours were Tangerine, Green Apple, Grape, Very Cherry, Lemon, Liquorice, Root Beer and Cream Soda. (Very Cherry remains a perennial favourite and for more than two decades has been the most popular bean.) Klein referred to his confectionary creations as Jelly Belly beans and to himself as Mr Jelly Belly.

By 1980, Klein had sold the Jelly Belly trademark to the
Herman Goelitz Company for $4.8 million and the
company's revenue doubled from $8 million to $16 million.
The bean quickly reached a celebrity zenith by confectionary
standards in the early 1980s, when President Ronald
Reagan's affectation for the sweets became the stuff of
legend. (Air Force One had special jelly-bean holders to
combat turbulence during flights, and Reagan even
surprised the astronauts of NASA's STS-7 mission with
Jelly Belly beans by having them put on the mission's
shuttle in 1983.) A portrait of Reagan, made from 10,000
Jelly Belly beans, hangs in his presidential library in Simi
Valley, California.

Today, Jelly Belly produces enough beans every year
to wrap around the Earth five times – something like
13 billion beans per year domestically with an additional
two billion per year internationally. According to Jelly
Belly official statistics, the company produces 136,000kg
(300,000lb) per day, 1,250,000 beans per hour and 1,680
beans per second. But it's their flavours that pique particular
interest and have vaulted them into a stratosphere-like level
of celebrity in the confectionary world.

The company boasts that Jelly Belly beans are made
with real ingredients, like 'fruit, peanut butter, and
coconut' and 'not the six or seven usually artificial and
often unidentifiable flavors' of regular jelly beans, as the
company emphasised in a 2008 *New York Times* profile.
These real ingredients give Jelly Belly jelly beans natural
flavour and, logic implies, natural flavour outranks artificial
flavour every time. Sarah Gencarelli, a confectionary
reviewer for CandyAddict.com, described these other jelly
beans with their lesser, limited artificial flavours to the
Times as 'the kind that were probably stuck together in
your grandma's candy dish'.

But Jelly Bellies are not your grandma's flavours.

Jelly Belly jelly beans come in Klein's original flavours, to be sure, but beans also come in flavours that range from Tabasco, to Bacon, to Margarita, to Egg Nog. Jelly Belly was the first company to debut a 'savoury' flavour candy – its signature Buttered Popcorn. Lychee-flavoured beans are available only via international orders (or at shops in Australia and China); jelly beans that taste like Green Tea can only be found on non-US sites like those in Greece or Germany. The now-discontinued Barbecue Banana flavour came in a box that proudly proclaimed, 'No Artificial Colours Or Flavours!'

At the turn of the twenty-first century, the Jelly Belly Candy Company upped its flavour game again by releasing a Harry Potter line of jelly beans and a plethora of 'unusual flavours' like Vomit, Earwax, Earthworm and Soap. To be fair, the Harry Potter line also included oddities that were, technically at least, based on actual foodstuffs like Sardine, Black Pepper, Grass, Horseradish, Spaghetti, Spinach, Sausage, Pickle and Bacon. (Charles Spence noted that the name of the flavour could heavily influence what we think a flavour tastes like – for example, the 'smelly cheese' and 'sweaty socks' share an odour chemical signature. One is perceived to be relatively pleasant; the other is the stuff of Jelly Belly's gross jelly bean line.) It's as if the flavours of Willy Wonka's three-course meal gum have come to life in a way that is more surreal than Roald Dahl could have ever imagined.

In 2008, the Jelly Belly Candy Company introduced BeanBoozled, a 'wild, risky adventure' where jelly beans looked 'normal' on the outside but had 'bizarre and gross' flavours on the inside. The gist was that consumers would be completely unable to tell what the flavour of a jelly bean would be – based on its colour – before popping it in their mouths for a quick chew. A white-coloured jelly bean, for example, could be Stinky Socks or Buttered Popcorn. A speckled one might be Birthday Cake or, perhaps, Dirty

Dishwater. The first edition debuted with Top Banana vs Pencil Shavings; Juicy Pear vs Booger; Liquorice vs Skunk Spray; Coconut vs Baby Wipes. The others could be Stink Bug, Dead Fish, Spoiled Milk, Toothpaste, Canned Dog Food and Mouldy Cheese. The list of repulsive flavours goes on and on. The game was a confectioner's way of asking customers if they feel lucky.

Playing BeanBoozled has, of course, turned into a genre of ironic YouTube performance art, and you can just picture a befuddled Gustav Goelitz stroking his handlebar moustache, outside his first Ye Olde Timey Candy Shoppe in Belleville, Illinois, trying to reconcile the idea of a company that started out inventing candy corn with a company whose website now boasts, 'JellyBelly.com is the best place to buy gross jelly beans, and we're constantly expanding our assortment to delight adventurous eaters!'

Fundamentally, the story of today's Jelly Belly is a story about flavour and all the chemistry, biology, history and culture that go into imagining and creating flavours, and then manufacturing sweets that carry those certain flavour expectations with them. Jelly Belly has built itself a rather artisanal foodie ethos, and how it markets the flavour of its beans is no small part of that story. 'We always start by sourcing the real thing. In the case of BeanBoozled, this is an interesting challenge,' Jelly Belly explained to me in an email interview. 'We have done things such as put used gym socks in a plastic bag for a couple of weeks or let milk spoil. The timeline for flavour development varies greatly by flavour. We constantly ask ourselves, did we get it as true to life as possible?'

But flavour is complex. And its history even more so. 'There are still luxury foods, like caviar and guinea hen, but there are no longer any luxury flavors,' a food manufacturer told Lucy Kavaler in her 1963 book, *The*

Artificial World Around Us. Kavaler's book, geared towards a young adult audience, follows how the world was remade by synthetic chemistry after the Second World War. With new, trendy twenty-first century flavours like Blue Raspberry and Acai, it would seem that new luxury flavours are being manufactured to take the place of those once held by nature. The excitement surrounding new chemistry-made flavours was much like that of laboratory-grown diamonds.

What makes a flavour natural? Real? Authentic? Biology. Biochemistry. Neurophysiology. History. Culture.

It's easy to think that flavour must be either 'natural' or 'artificial'. Certainly, the minutiae of contemporary food labelling, ingredient lists and legal regulations emphasises this distinction – natural and artificial mean very particular things in specifically legal circumstances, regardless of an individual flavour's history. This dichotomy is overtly and covertly translated into the consumer ethos that flavour is either real, or that it is fake and that that 'natural' is 'good' and 'artificial' is 'bad'. Fundamentally, flavour is about decisions and how the sensation of taste moves through the human experience.

But such rigid categories don't let flavour evolve. Like so many things, flavour isn't easily divided into two distinctly separate camps, let alone easily categorised as 'fake or not'. If history is any indication, what flavours we think are real and fake will change – and change again – in the decades to come.

CHAPTER FIVE

Taking a Look Through Walrus Cam

Every summer, thousands of Pacific walruses swarm the dark grey Arctic beaches of seven craggy islands in Alaska's northern Bristol Bay. Every year, from May through to August, tens of tons of these tusked mammals heave their considerable bulk out of the ocean to sun themselves, groom themselves (backscratching is a perennially favourite pastime), and in general rest from their near-constant foraging for food – a set of walrus behaviours that biologists refer to as 'hauling out'. And every summer, thanks to Explore.org's Walrus Cam, tens of thousands of humans watch them do it.

In 1960, this small Alaskan archipelago was designated as the Walrus Islands State Game Sanctuary to protect walrus haulout areas by carefully controlling the area's land and sea traffic. There aren't any incorporated cities in the Bristol Bay borough, and as of the 2010 census it claimed only 997 residents spread out over almost 2,330km² (900mi²). Consequently, the sanctuary, surrounded by water and wilderness, is pretty much inaccessible to anyone other than field biologists and the most ardently determined of wildlife photographers. Though low in human residents, Round Island is the largest terrestrial walrus haulout site in the world, offering beachfront real estate for the walruses' summer repose in their migratory journey between Russia and Alaska, along the Bering Strait. During the summer months, thousands of gregarious walruses bask along the shorelines of Point Lay, day after summer day, crammed side-by-side, tusk to whiskered jowl, like one-ton sardines.

Formally known as *Odobenus rosmarus*, the scientific name for walruses roughly translates from its Scandinavian root as 'tooth-walking sea horse', and that's not necessarily an inaccurate description of these pinnipeds. Walruses use their massive canine tusks to lever themselves in and out of the water on to ice floes during the winter months, when they are not near land. Palaeontologists can trace the evolutionary lineage of the modern walrus back 10 million years, as evidenced by a fossil of a tuskless ancestor found off the coast of Japan. Other fossils from the Pleistocene show that walruses and walrus-like ancestors have inhabited cold environs for the last million years, following global shifts in climate to stay in icy areas. Although modern, extant walruses are ecologically tied to their icy Arctic environments and have been for millennia, there is certainly a seasonal, annual dynamic to their behaviour.

That seasonal migration cycle is what brings the walruses to their annual haulout on the Round Island beach. Although the number of walruses that beach themselves varies day by day and year by year, the Alaska Department of Fish and Game reported that record numbers of walruses – 14,000, 10,000 and 30,000 – were counted on the Round Island beach in various daily censuses. CNN even reported that more than 35,000 were observed on 27 September 2014, based on aerial data from the National Oceanic and Atmospheric Administration. The summer of 2015 not only saw colossal numbers of walruses at Point Lay beach; it was also the summer that Walrus Cam turned back on.

Among wildlife biologists and conservationists, the Walrus Islands State Game Sanctuary offers space to study the delicate northern ecosystem, as well as opportunities for public education and outreach about climate change and its effects on the surrounding environment. Ten years earlier, the Alaska Department of Fish and Game had established a live webcam feed with a camera pointed towards the walrus haulout on the beach. While the webcam was immensely popular with internet wildlife enthusiasts, the department lacked the technological infrastructure and funds to keep the livefeed going and the website that hosted the feed regularly crashed. (In 2005, YouTube was still in its internet infancy and wasn't the go-to resource for live video streaming that it would become a decade later; consequently, Fish and Game had attempted to host the feeds itself.) However, as the department learned, tens of thousands of humans tuning in to watch tens of thousands of walruses pushed the live-streaming capabilities of Walrus Cam well beyond what the department or its IT team could sustain, and Walrus Cam went into deep hibernation for the next 10 years.

Walrus Cam eventually turned back on due to the philanthropic efforts of Explore.org, a multimedia organisation that develops and finances projects devoted to exploration, especially projects that showcase wildlife and nature. (One of Explore.org's best-known projects is Bear Cam, a live-streaming feed trained on a small waterfall where Alaskan Brown Bears fish for salmon in the Canadian Brooks Falls area.) With Explore.org's financial backing, armchair wildlife enthusiasts have watched Pacific walruses haulout every summer since 2015, all from the comfort of their personal internet connections. The Alaska Department of Fish and Game maintains the two cameras and feeds over the island's summer season. The popularity of Walrus Cam is indisputable – in 2016 it drew more than 1.3 million views on various video-streaming and sharing sites.

Today, people watch the mottled pink and brown walruses of Walrus Cam's world slap each other with their flippers, tusk neighbours over space, and lurch in and out of the fixed camera frame. Walruses shuffle up and down their

beach, and even resting walruses offer drama to the feed as
someone puts their flipper in someone else's eyes, someone
has an itch that needs scratching, or a walrus gets a bit too
close to someone else and the tusks are flashed to defend
personal space. Waves crash in the background, tides come
in, walruses waddle back out to the sea and, if viewers
listen closely enough, there is the occasional bird squawk or
walrus grunt piped through their internet connection. The
bonhomie is palpable.

On its most basic level, Walrus Cam shows the everyday
lives of everyday walruses going about the business of doing
their everyday walrus things. With Walrus Cam, there isn't
any script. There isn't an authoritative narrator to give
viewers a story to go with what they're watching. There
isn't any scientist explaining walrus behaviour to the
non-expert viewer to help them understand subtleties and
dynamics that an amateur nature watcher might miss.
('Over here, we see two walruses striving for dominance to
be the alpha in their huddle.' 'That walrus is so tired from
foraging that he hasn't joined the rest on the beach.' 'That's
not actually a walrus, that's a rock with a bird on top of it.')
There isn't any camera crew wading through animals to
get the best footage. It is a completely unscripted way to
watch animals in nature.

Walrus Cam, as its watchers can attest, fosters a feeling
of connection between its pinnipeds and its humans.
(Reader, I named a trio of walruses Brad, Thad and Chad
while watching the Point Lay Beach feed one afternoon.)
Without the narrative and storytelling that goes into
formal wildlife and nature documentaries, Walrus Cam
watchers are left to project what they see walruses doing
based on whatever they happen to already know about
walruses – and whatever factoids they glean while toggling
between Walrus Cam, Wikipedia and Explore.org's forum
questions.

Above: The skull of *Microraptor* under white light (top) and Laser-Stimulated Fluorescence (bottom). The colour differences indicate changes in mineralogy, which means that this is a composite specimen.

Below: Two examples of Johann Beringer's 'Lying Stones'. OUMNH T.23 (left) depicts slug-like creatures and OUMNH T.22 (right) depicts arrows and boomerang-like symbols.

Left: William Henry Ireland's purported letter from William Shakespeare to Anne Hathaway was one of his more popular forgeries.

The Letter runs line by line :—

Dearest Anna,

As thou haste alwaye founde mee toe mye Worde moste trewe, soe thou shalle see I have stryctlye kepte mye promyse. I praye you perfume thys mye poore Locke withe thye balmye Kysses, forre thenne indeede shalle Kinges themmeselves bowe ande paye homage toe itte. I doe assure thee no rude hande hathe knottedde itte; thye Willys alone hathe done the worke. Neytherre the gyldedde bawble that envyronnes the heade of Majestye, noe, norre honoures moaste weyghtye, wulde give mee halfe the joye as dydde thysse mye lyttle worke forre thee. The feelinge thatte didde nearest approache untoe itte was thatte whiche commethe nyghste untoe God, meeke and gentle Charytye; forre thatte virrtue, O Anna, doe I love, doe I cheryshe thee inne mye hearte; forre thou arte as a talle cedarre stretchynge forthe its branches ande succourynge smalle Plants fromme nyppinge Winterre orre the boysterouse Wyndes. ffarewelle, toe Morrowe bye tymes I wille see thee; tille thenne, Adewe, sweete Love.

Thyne everre,
Wᴹ. SHAKSPEARE.

A LETTER, FORGED BY IRELAND, PURPORTING TO BE WRITTEN BY SHAKSPEARE TO ANNE HATHAWAY.

Below: The Grolier Codex, a Maya codex made of fig-bark paper, contains a Venus almanac and dates between AD 1021 and 1154. The document is currently held by the Museo Nacional de Antropología in Mexico City.

Left: A nineteenth-century painting by the Spanish Forger using oil on wood. Notice the cracks that help 'age' the painting, and the Spanish Forger's signature 'tells' – the bow mouths, turned-out feet, and daring décolletage.

Left: A manuscript page painted by the Spanish Forger on parchment. It is a full-page illustration of the Visitation, and was painted on the back of an authentic 15th-century antiphonal leaf (below). Notice the cutesy castles and lollipop-like trees – both Spanish Forger 'tells.'

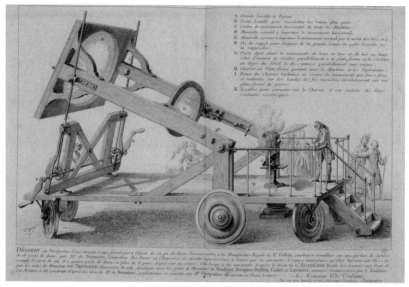

Above: Antoine Lavoisier (in goggles) operates his solar furnace. Lavoisier used this apparatus to burn diamonds in glass jars; tests of the products of combustion led him to show that diamonds are comprised solely of carbon.

Below: Herbert Strong examines a laboratory-grown diamond, 1970.

Below: The research team from General Electric responsible for creating laboratory-grown diamonds, 1955.

Left: Stereograph of leafy banana trees housed in the Floral Hall at the Philadelphia Centennial in 1876.

Below: Employee of the Bureau of Chemistry's Phytochemical Laboratory using a distillation apparatus to concentrate the odorous constituents of apples and peaches.

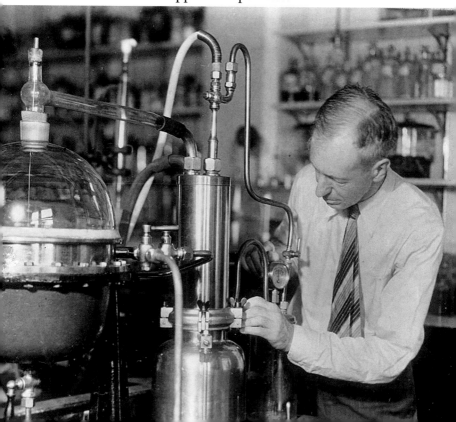

Above: Big Blue as she hangs in the Beaty Biodiversity Museum, University of British Columbia.

Below: Hanging model of the Blue Whale in the Irma and Paul Milstein Family Hall of Ocean Life. Newspaper insert from the front page of the *Columbus Dispatch*, 8 March 1881.

Left: Walruses as seen via Walrus Cam at Round Island, Alaska, 2018.

Left: Disney's 1958 film *White Wilderness* showed Brown Lemmings committing 'mass suicide' by jumping off a cliff, even though this was not a natural behaviour. A later investigation found this scene was completely staged.

Below: Gilles Tosello's workshop in Toulouse, creating parts of the Caverne du Pont d'Arc.

Above: Painting walls in Caverne du Pont d'Arc to replicate those found in Chauvet Cave. Image extracted from the 2015 documentary film *Les génies de la grotte Chauvet*.

Below: *Peckham Rock* by Banksy. This was a rogue installation placed surreptitiously in the British Museum by the artist in 2005. The piece was then loaned (back) to the museum as part of the I Object! exhibit in 2018.

Walrus Cam promises a nature – the Real Thing – that you would see, if you were only there.

★ ★ ★

Live-streaming nature through feeds like Walrus Cam is a uniquely twenty-first-century way to watch walruses, and with it come uniquely twenty-first-century expectations of realness, reality and authenticity about the nature that you are watching. It's both a reaction and an answer to decades of wildlife documentaries, and the audience expectations that have come with them. Wildlife documentaries have established themselves as *the* authoritative way to watch filmed nature, and that authority carries a lot of cultural cachet, as well as a price. Over the last six decades, wildlife films have come to be built out of expertise, whereas live-streaming tends towards unscripted work. One is driven by blockbuster sales, the other struggles to keep the cameras on. Both, however, hinge on audiences' belief that what they're seeing is authentic.

Although Walrus Cam brings in tens of thousands of viewers to the wildlife-laden beaches, it only does so during the summer months, when the walruses haulout on the beach, which also happens to be when audiences watch. The rest of the year the cameras are turned off and the footage for Walrus Cam via Explore.org simply cycles through 'best of' clips for anyone who loads the feed. Wildlife documentaries, however, aren't constrained by this sort of seasonality or logistical limitations. Wildlife films, the reasoning goes, exist for people to be able to watch them whenever and however they want, as long as they've paid to do so.

This on-demand availability reflects the cost and audience expectations that come with it. Bringing nature to viewers doesn't come cheaply, and as audience expectations for wildlife documentaries have risen, so have their production costs. Explore.org might be able to get away with the relatively low costs associated with live-streaming in an Arctic ecosystem, but that laissez-faire approach doesn't work in the competitive and highly lucrative business of blockbuster personality-driven documentaries.

The cost of bringing wildlife films to audiences can be measured in time, effort and outlay of cash, and twenty-first-century blockbuster documentaries can boast spending all of these. When *Blue Planet II* was released in October 2017, it was said to have cost the BBC's Natural History Unit something like $25 million to film, produce and distribute. Estimates put the landmark, genre-defining 2006 *Planet Earth* at costing around $2–2.2 million per episode, over $11 million in total. Both National Geographic and the Discovery Channel allocate at least $400,000 per episode. *March of the Penguins* (distributed by Warner Pictures, 2005) boasted a budget of $8 million. These are staggering sums.

More than just money, however, there is a considerable outlay in the time, effort and expertise devoted to securing the footage for the films. *Blue Planet II* (2017), for example, was able to boast four years of production running something like 125 expeditions, more than 6,000 hours of underwater dive footage and 1,000 hours of filming in submersibles. The sharp, crisp, spectacular footage in *Planet Earth II* (2016) is the result of incremental, steady improvements to camera technology made over the decade since the original filming of *Planet Earth* (2006). Audiences know that the studios have spent inordinate amounts of time and effort to make the films because the documentaries tell them that they have. All twenty-first-century landmark wildlife films feature a 'behind the scenes' tour that shows viewers the time and creativity needed to create the films in the first place. Such costs in effort and expertise have come to define the film genre as much as the spectacular footage itself.

The pay-off for studios is the millions and millions of viewers who stream, watch and scrutinise filmed wildlife and pay for the privilege of doing so. *Planet Earth* (2006), for example, experienced record-breaking DVD sales and the series was ultimately broadcast in 130 countries with a total of 100 million viewers. Five years later, 48 per cent of the UK population watched at least 15 minutes of *Frozen Planet* (2011). When *Blue Planet II* aired in autumn 2017, the first episode in the series was watched by 14 million people, and it quickly became the most-watched programme on UK television for the year and the most popular natural history series in 15 years, beating the previous record holder, *Planet Earth II*, which could boast a 'mere' 13.1 million viewers. *Planet Earth II* was in fact able to brag that it had 1.7 million more viewers than any episode of celebrity dance-off *Strictly Come Dancing*. The sheer command of audience numbers is stunning.

Audiences expect exquisite footage from a traditional wildlife documentary, pulled together through expert filmmaking and emotional storytelling – all of which stick with audiences long after the show has finished. 'A whole generation of wildlife lovers was shaped by films they saw decades ago,' wildlife film producer Chris Palmer explains in his memoir, *Shooting the Wild: An Insider's Account of Making Movies in the Animal Kingdom*. 'Whether it was *Death of a Legend*, Jacques Cousteau's *The Silent World*, Marlin Perkins's *Wild Kingdom*, or a movie in the Disney *True-Life Adventures* series, images on a screen sparked a fascination with nature.'

On its most basic level, the entire genre of wildlife documentaries concentrates on introducing animals and plants in their natural habitats to audiences – offering stories, spectacle and science to pique interest. Ever since

Disney released the first of its *True-Life Adventures* in 1948, audiences have come to expect certain story elements from the wildlife films that they're watching, and twenty-first-century documentaries deliver on those historical assumptions about animal storytelling.

Cynics might argue that the ballooning costs of wildlife documentaries in the twenty-first century are simply a reflection of the capitalist exploitation of nature. They point out that because audiences have been conditioned to watch for great footage, filmmakers constantly need to look for the next big thing to keep pace, essentially commodifying the very animals and environments that they seek to document. 'In such broadcast climate,' science and technology studies scholar Eleanor Louson argues in an article about wildlife documentaries for the journal *Science in Context*, 'it is difficult for film producers to make money unless they first spend it, precisely to obtain the spectacular visuals.' Expectations about animals and how we see them, then, have become requirements for filmmakers to satisfy their audiences. As audiences, we have agreed to suspend our disbelief and to have faith in the authenticity of filmmakers' dramatisation of wildlife – we believe that what we are watching is genuine, true and real.

Curiously enough, what makes filmed footage 'real nature' – and the best way to convey that realness to audiences – changes over time. In other words, what was real for a Disney wildlife film in 1948 might not be real for the BBC in 2018. And if the history of documentary films is any indication, what is real for 2018 will change in the next 50 years. This flexible reality within wildlife filming begs the question of what's at stake and why it matters. While everyone, filmmakers, audiences and studios, might agree that these things – reality, truth and realness – underscore and legitimise what makes a good wildlife documentary, just what makes something authentic is a bit tricky to pin down.

First and foremost, according to Jeffery Bosall, a media theorist and producer at the BBC Natural History Unit from 1957 to 1987, two things delineate an authentic wildlife documentary: namely, that audiences are not deceived, and animals are not harmed in the making of the film. (Indeed, you would be hard-pressed to find a reputable wildlife film producer who thought that deceiving audiences and harming animals was acceptable.) Bosall later extended these criteria, advocating for on-screen disclaimers about what parts of the film were staged. Bosall's premises are easy enough to agree with, but often difficult, as it turns out, to consistently translate into practice. It's hard – it's really, really hard – to tell a story about animals that might be uncooperative, film a documentary under conditions that aren't conducive to getting necessary footage, or film something that might cause distress to animals. You might be violating the second, for example, by upholding the first.

And this is where filmmakers are engaged in a delicate tango of blending art and artifice into the world of storytelling in wildlife films – the balance of the two is under constant negotiation and subject to ever-changing and ever-evolving expectations.

★ ★ ★

Just as audiences expect authentic, ethical wildlife films with spectacular footage, they also expect those films to carry a compelling story. And the story of how those stories have been told has a history all its own.

In the late 1940s, the Walt Disney Company launched its influential and innovative *True-Life Adventure* films, kicking off the series with the release of *Seal Island* in 1948. Walt Disney himself was particularly intrigued with the idea of an Alaska travelogue, based on stories he had heard from servicemen who had been stationed in Alaska during the Second World War. He pitched the idea to one of his company's minions that Alaska was, in his words, 'our last frontier, the last undeveloped place in the United States. We should have some photographers up there. Look into it.'

Disney hired filmographers Elma and Alfred Milotte to head up to Alaska and see what they could shoot over the year of 1947. The Milottes shipped 30,480m (100,000ft) of film back to Walt Disney Studios, much to the cost-conscious consternation of Disney's financial team. Their footage focused on people, animals and environments. (Walt's older brother Roy finally asked Ben Sharpsteen, the director of *Pinocchio*, 'What is Walt going to do with all that Alaska stuff?' 'Perhaps make some sort of glorified travelogue out of it?' was Sharpsteen's guess.) In fact, Sharpsteen himself would eventually direct that 'glorified travelogue'.

Although Walt was intrigued by the Milotte footage, it wasn't until he took his 10-year-old daughter up to Alaska that he could see how to build a story out of grand views and breathtaking scenery. That was the trip during which Walt found the seals. They were cute, personable, curious and utterly unfamiliar to Disney's audiences. Perhaps with enough backstory they could be the non-fiction counterparts to Bambi and other Disney animals, or so the internal logic went.

'Why don't we take what we have and build a story around the life cycle of the seals? Focus on them – don't show any humans at all,' Walt suggested. 'We'll plan this for a theatrical release, but don't worry about the length. Make it just as long as it needs to be so you can tell the story of the seals.'

The short film ran just 27 minutes in length. Walt was so thrilled with the result that he announced that *Seal Island* was the first in an entire series about nature. (Although he had no ideas about what else might be part of that series.)

Some studio executives balked at the entire notion of a film built around a group of seals, and Disney decided to demonstrate audience interest by taking the film directly to audiences. In December 1948, Walt persuaded Albert Levoy – who operated the Crown Theater in Pasadena – to book *Seal Island* in conjunction with a regular-length feature. Five thousand questionnaires were handed out to audiences and the results were outstanding: they preferred *Seal Island* to the feature itself. Not only did they prefer it to the feature film, they wanted more like it.

Seal Island won an Academy Award in 1949 for Best Short Subject, Two Reel. The opening credits for *Seal Island* boasted that the film was 'unrehearsed and unstaged', and it opened with animated sea mist brushed away by an animated paintbrush, revealing a live-action panorama behind the art. The narrator announced that on Seal Island, such that it was, 'nature plays out one of her greatest dramas ...' and that what will follow is 'theater for the spectacle'. In other words, *Seal Island* was entertainment, yes, but it was also 'authentic'.

Comprising 10 short and four feature-length films released in 1948–1960, the *True-Life Adventures* series defined the genre of wildlife films, according to media historian Morgan Richards. The series synthesised the then-familiar tropes of safari, scientific-education and ethology films that audiences were more than familiar with, and blended each with Disney's already popular cartoon, comedy and Western genres. 'Disney's break-through lay in its ability to dramatize the natural world and bring wild animals and nature to life using full colour cinematography and lavish musical scores,' Richards observed in a contributed chapter to *Environmental Conflict and the Media*. 'It was their glossy finish and sense of drama, more than anything else, which essentially distinguished Disney's films from other wildlife fare and gave them a commercial edge, an edge that was further honed through Disney's monopoly over distribution.'

True-Life Adventures were set in all sorts of American wilderness. From *Seal Island*'s Arctic Alaska to Florida's Everglades, from Arizona's deserts to Oregon's prairies, the animals and their habitats enchanted audiences. The films followed all sorts of charismatic animals, and in 1953 the release of *Bear Country* received a glowing review in *The New York Times*. 'As for "Bear Country," it follows in the excellent series of nature films that have been produced by the Disney studio ...' Bosley Crowther raved in his review. 'This one studies the environment and habits of American wild bears to reveal both the sturdiness and cunning of these amiable-looking animals. The most amusing sequence in the picture is a montage of bears scratching themselves, done to a deft bolero rhythm. It's a case of trick editing, but it's fun.' (Fourteen years later, audiences would re-encounter that very bear back-scratching in choreography of Baloo the bear, singing 'The Bear Necessities' in Disney's animated film *The Jungle Book*.) This was the American wilderness as encountered, explained, packaged and anthropomorphised through Disney.

At its core, the success of *True-Life Adventures* was the sentimental and sanitised vision of nature that it offered its audiences. Nature might not be harmonious, Disney films went, but it could be neatly moralised into an easily digestible story for its audience. In *Seal Island*, bull seals are called 'beach masters' and females are 'brides' – at one point, the musical score actually kicks into a spritely rendition of 'Here Comes The Bride!'

According to philosopher Derek Bousé in *Wildlife Films*, 'The portrayal in wildlife films of the animal's family and social relations present a kind of vast Rorschach pattern in which culturally preferred notions of masculinity, femininity, romantic love, monogamous marriage, responsible parenting, communal spirit … can all be read.' In the decades after *True-Life Adventures*, successful nature films followed Disney's formula: feature classic megafauna (big cats, primates, elephants, predators); situate animals in 'primeval wilderness'; maintain dramatic and suspenseful storylines; and, in general, avoid science, politics and anything controversial.

Disney was hardly the first to sculpt nature into parables for general living. From Aesop's counselling that 'slow and steady wins the race' in his fable 'The Tortoise and the Hare', to Jack London's twentieth-century incarnation of pure nature in *White Fang*, humans have been making animals in their image for millennia. What made *True-Life Adventures* different, however, was that it set the stage for every wildlife film produced after it.

Although such morally steered megafauna was a decades-long sure bet for successful storytelling, several film studios began to subtly challenge the classic Disney formula in the latter half of the twentieth century. While studios like Discovery and National Geographic continued to create Disney-esque films (called 'blue-chip documentaries' in the industry), other studios – in particular the BBC – began to explore other ways of introducing audiences to nature, and of negotiating the boundary between humans and the environment that such filmed entertainment offered. Several of these studios offered a neat, crisp lecture-type of show, where a tweed-wearing naturalist in a room back-dropped with books and stuffed specimens would take viewers through the lives of various animals. This is the era that iconised the dulcet narrative tones of Sir David Attenborough in the BBC's *Life on Earth* series, which kicked off in 1979.

In October 1996, BBC1 introduced a new wildlife series that featured the lives of big cats in East Africa – *Big Cat Diary*. The footage used in the series was shaky, incomplete and less than perfectly polished compared with what Disney had produced over the decades. The idea behind *Big Cat Diary* was that audiences would watch the crew film the everyday lives of big cats (a lion pride, a cheetah family and a leopard family) on a Kenyan game reserve. 'This is the Masai Mara in Kenya, one of the best places on Earth for watching wildlife,' *Big Cat Diary* presenter Simon King stated. 'But in this series we are going to be looking at it in an entirely new way. Over the next six weeks we are going to be following in intimate detail the lives of Africa's big cats, sending back a weekly report, a diary of their hardships and good fortunes as they happen.'

This was reality television come to nature. This was raw, this was edgy and this was *real*. Even the title sequence that opened each of the 76 episodes illustrated this new aesthetic. Lions pawed at each other, cheetahs chased down their prey, iconic animals came in and out of frame, and camera crews meandered down bumpy grassland roads – all against the backdrop of Masai vocals and drum beats. The final scene cut to the *Big Cat Diary* logo on the front door of one of the field jeeps, spattered with mud. It was as if the crew had just stepped off its savannah adventures to stop and show you footage of all the amazing things that big cats do. It was, in film industry parlance, a docusoap.

Although it is easy to dismiss *Big Cat Diary* as simply reality television with lions in lieu of Kardashians, the BBC1 series forced a new aesthetic on viewers about what it meant to watch nature. It required viewers to – consciously or not – decide if what they were seeing was true in a way that earlier, classic *True-Life*-esque films did not. Earlier wildlife films were true because the story and narrator led audiences to think that they were true. In *Big Cat Diary* the film was true because audiences could see it being filmed and cut together. Even if the cats spent days on end doing 'uninteresting' things, well, that was what the cats did, so that's what audiences would see.

To that end, the narrators for *Big Cat Diary* weren't omniscient, removed figures who talked over footage and explained nature to audiences from a comfortable distance of the third person. (Disney's early narrators, like Rex Allen and Winston Hibler, quickly became iconic mid-century Americana – voices remembered and associated with the wilderness they talked about.) *Big Cat Diary* introduced a nature programme in which the narration and storytelling derive from a personality presenter. Personality-driven wildlife films become immensely popular across a plethora of late-twentieth- and early twenty-first-century series, (Steve Irwin's *The Crocodile Hunter Diaries*, for example.) But this type of presentation was – is – a double-edged sword. It means that the film, at its core, isn't about the animals; it's about the presenter. The presenter has to carry the story, because the animals aren't scripted to do so. Critics scorned the lot of them as 'brash showboating'. And these reality-driven films blurred the line between real and not in ways that earlier films had not.

Big Cat Diary offered an authenticity – or, at least, a perceived authenticity – to viewers because audiences could watch how the films were made. Audiences could see the difficulties in getting certain shots, and could observe the long periods of waiting that come with making a wildlife film. This, reality docusoaps promised, was real nature the way nature intended it, and it was true because you were there to see it, thanks to BBC1. While earlier wildlife films deliberately took humans out of the picture, in reality-docusoap films humans were there, front and centre. The watcher had become the watched – the crew and filmmakers were openly acknowledged to be part of the story of wildlife films.

'In *Big Cat Diary*, the lives of wild animals were being framed and edited in a new way,' Morgan Richards offers in a scholarly reivew about docusoaps. 'Viewers were being offered a vision of wildlife that was intimate, personal, and self-consciously rough edged.'

★ ★ ★

Successful wildlife documentaries have continued to advance in their craft and technology, drawing heavily on the historical legacy of Disney combined with the cachet of audience buy-in for wildlife docusoaps. Today, efficacious wildlife films are blockbuster extravaganzas, with all of the plaudits and expectations that go with hitting it big. Twenty-first-century documentaries like *Planet Earth I* and *II*, *The Blue Planet I* and *II* and *Frozen Planet* – to name a few – boast huge budgets, big-name stars, dramatic storylines, behind-the-scenes footage, award-winning scores, sequels and even spoilers.

In the media world, a blockbuster is more than just a financially successful film. It's a production that has mass marketing behind it. It has stuffed animals, soundtracks and more behind-the-scenes footage than a non-blockbuster. (Several colleagues swear by the *Blue Planet* soundtrack as their favourite film soundtrack.) Studios spend inordinate amounts of money filming, writing, producing and distributing the film – why not find every way possible to squeeze out more money from audiences?

To that end, blockbuster films trade on the cachet of spectacle and are built on the expectations that what audiences are seeing is, well, spectacular. In a world dominated by YouTube, why would audiences pay to watch animals on film when they can do it for free? Paying audiences won't tolerate grainy footage or boring narratives – blockbuster audiences want to see something that they couldn't see anywhere else and be told a story that they haven't heard before. Blockbuster nature films fit that bill.

The blockbuster model for wildlife films can be traced to Alastair Fothergill, his time as director of the Natural History Unit at the BBC, and the release of the original *The Blue Planet* series in 2001. Fothergill convinced the studio to take the series, costing the BBC £6.7 million ($10 million) and seven years of time. This sort of investment of both time and money was unprecedented in the world of wildlife films, but Fothergill's gamble paid off – more than 12 million people tuned in to the series, and it sold in over 50 countries, regularly achieved an audience share of 30 per cent, and won both Emmy and BAFTA awards. In 2003, the series was recut and repackaged into a single 90-minute self-contained film, then the entire thing went on theatrical tour (*The Blue Planet Live!*) in 2006–2008. This was truly spectacle.

Spectacle, however, is complicated. Nature has been viewed, watched and observed as spectacle for centuries. Long before Disney was filming seals or *Planet Earth II* offered viewers the epic saga of an iguana outrunning a writhing ball of racer snakes, enthusiasts – amateur and professional alike – were connected to the natural world via the feelings of awe and wonder fostered through natural-history dioramas, cabinets of curiosities, and vividly painted landscapes and specimens. Artists used vibrant colours, bookbinders created large-format folios to outsize regular books, taxidermists stuffed only the most regal of dead animals and collectors vied for the most attention-grabbing of specimens.

Nature, the logic went, was meant to inspire majesty and wonder, and the medium to do so must be equally matched. Contemporary wildlife documentaries continue that centuries-old ethos of showing audiences nature as a wondrous and awe-inspiring thing. With contemporary documentaries, then, the power and attraction of spectacle continue. Here, viewers are spectators – active participants in the very process of looking at nature – thus coming full circle back to the spectacle that they're watching. 'The spectacular footage present in contemporary wildlife filmmaking is continuous with these traditional forms of visual natural history,' Eleanor Louson suggests in her study of contemporary wildlife filmmaking, 'each designed in order to elicit wonder and awe, and reiterate longstanding trends of visual reasoning, collection and display.'

While *Planet Earth* codified the wildlife film as a blockbuster, there have been aches and pains along the way in making these landmark series 'good'. For example, although few would dispute the visual spectacle of these films, the balance between spectacle, science and storytelling has been slowly tipping in favour of more drama in recent documentaries. Audience anticipation and reaction, queued through the visual pictures set out in front of it, would seem to trump building a nuanced, sophisticated story. 'Nature documentaries have been limited by the same technology that makes them so compelling. When you can get beautiful, high-definition, slow-motion, ground-level shots of an animal, it's not enough to just show it and start talking. You need build-up,' science journalist Ed Yong suggests in *The Atlantic*. 'The storytelling language of wildlife documentaries has become more cinematic, and every vignette becomes longer. That necessarily reduces the amount of material you can get through in a given hour.'

In addition to spectacular footage, contemporary audiences assume – expect, even – that wildlife documentaries play an important role in public understanding of science and environmental issues and, certainly, these films carry an awful lot of cultural cachet to legitimise conservation and environmental narratives.

However, wildlife films have not always gathered their authority from the scientific community, and most of the earliest twentieth-century films were firmly centred on the business of entertaining audiences and the profits that came with that. When the final episode of the BBC's *Frozen*

Planet series discussed the environmental impacts and reality of climate change, the Discovery channel entertained the idea of not airing that 'environmental' episode to American audiences, who by all accounts would be more reticent to watch it than their British counterparts. Discovery, it would seem, just wanted polar bears without the uncomfortable science that goes into understanding the changes in polar bear ecosystems and the role humans play in those changes. Amid international outrage, Discovery relented and aired the 'climate-change' episode to American audiences. The outcome? Discovery made money and the Arctic continues to melt – so business as usual.

Introducing environmental science is, interestingly enough, a rather recent and somewhat still-fraught way to frame storytelling about nature. In other words, we're a long way from folksy naturalists explaining how animals live in their environments, or perhaps even a long way from the nominal educational value of *True-Life Adventures*. As each wildlife blockbuster tries to outdo its predecessor, the visuals remain spectacular, but it becomes easier for the story – and the science – to become more obscure. It's almost as if it's become spectacle for the sake of spectacle. Many wildlife film enthusiasts argue that the most recent series released by the BBC's Natural History Unit has struck a better balance of message and spectacle in *Blue Planet II*. 'Where previous series felt like they sacrificed the storytelling craft and educational density for technical wizardry and emotional punch *Blue Planet II* finally marries all of that together,' Yong stated in his review of *Blue Planet II* in *The Atlantic*.

We might be watching the wildlife-film pendulum start to swing to a place where spectacle in and of itself no longer meets all audience expectations.

★ ★ ★

All of this brings us back, however, to the question of how and why we believe what we watch. The very success of wildlife films hinges on their ability to show audiences the spectacular reality of nature – aka the authenticity of nature and all of its wildlife trappings. We believe that wildlife films are true, that we're watching something with a little more cultural credibility than grainy Bigfoot footage. We believe wildlife documentaries are true because we're expressly told that they are, or we believe it because we believe the stories around the films that result in what we're watching. We think that what we're seeing is real. And if it's real, then it must be authentic. Disney, remember, expressly claimed that its *True-Life Adventures* were 'completely authentic, unstaged and unrehearsed'.

The question of authenticity in wildlife films circles back to arguments about wildlife film ethics and best practices, which brings us, again, to Jeffery Bosall's argument that an authentic wildlife documentary doesn't deceive audiences and doesn't harm animals – and it's easy to think that simply following these two rules of best practice automatically translates into an authentic documentary. Fundamentally, however, the relationship between filmmaker and audience is built on trust – if audiences feel manipulated or deceived about what they're watching, the documentary loses its claim to authenticity. 'The history of wildlife filmmaking,' Eleanor Louson suggests 'is like that of the discipline of natural history itself – one of tension between authenticity and artifice in the display of nature.' But between these two pillars of best practices is a lot of negligible grey space filled with tricks of stagecraft and storytelling. It's walking this line of artifice and authenticity that makes wildlife documentaries such a curious way to view real animals in their real wildernesses.

On the artifice side of wildlife documentaries, there is a plethora of ways in which filmmakers can deceive their audiences. For example, if filmmakers introduced animals that don't normally interact (say, a walrus and a wildebeest), that wouldn't be natural. Or if they encouraged animals into a certain sort of behaviour (for instance forcing a confrontation between predators or scaring a bird off a nest), this would be real, but unnatural. Also considered egregiously fake? Sentimentalising, exaggerating or overdramatising animals. These are all ways that filmmakers build an unnatural story out of real parts. On the other hand, certain kinds of artifice are necessary to create an ethical wildlife documentary. To procure certain kinds of footage, filmmakers will often splice in footage of tame or captive animals from wildlife farms or zoos to illustrate part of an animal's story. Digitally manipulating footage after it is shot can help to cover less than ideal materials. CGI is at the burgeoning cusp of being able to help cut filmmaking costs, with the possibility of computer-aided wizardry being employed for future documentaries.

Staging a scene for a wildlife documentary offers a way for filmmakers to alleviate danger to themselves and animals. In many instances, it enables them to capture certain aspects of behaviour that wouldn't be available to film under 'normal' circumstances. Using everything from tame or captured animals at game farms to digitally altering footage is part of the trade-offs filmmakers use to mitigate their presence in the wilderness. There's a bit of an irony to the practice. The end documentary is a real narrative with scientific authority – the means and methods might employ some artifice and fakery to get there.

'I try to approach these complex issues in both practical and ethical terms,' wildlife filmmaker Chris Palmer said of his own work in his memoir. 'Let's say I'm trying to film an isolated detail of animal behaviour such as lobsters spawning. Following a noncaptive lobster around underwater, waiting for the right moment and right light, wouldn't be practical, so I would film lobsters spawning in a tank. With larger animals such as bears, ethical and safety issues become paramount: it's dangerous to get too close and imprudent to habituate wild bears to humans, so it makes sense to use captive, tame bears. It's a paradox, but captive animals make responsible wildlife films possible – so long as certain principles are honored.'

'Filmmakers tend to justify their staging practices by appealing to the underlying reality, educational value, or scientific truth of their footage,' Eleanor Louson expands on the principle that Palmer describes. 'Their justifications generally rely on arguments that appropriate "natural history artifice" was employed to generate an experience for viewers that is "more real" than could be found by viewers in real life, or to produce footage that, for reasons of cost, pragmatism, or efficiency, could not have been obtained otherwise.'

But this is a slippery slope of a way to tell stories and make films, because it's using artifice to achieve a narrative arc in its story. It's easy for audiences to feel deceived or cheated. When a documentary crosses the line – and sacrifices authenticity through artifice – the public backlash is vehement and impressive. One of the most-cited examples of ethically egregious fakery in nature films is the infamous 'lemming scene' in Walt Disney's 1958 *White Wilderness*, part of the *True-Life Adventures*. *White Wilderness* filmed everything from walruses to polar bears to narwhals. But it was footage of thousands of tiny lemmings swarming over a cliff's edge into the Arctic Ocean, with the implication that they were committing suicide on their migration path, which has really left its mark in the documentary film world as well as in popular culture. It looked as if the lemmings were simply jumping off of a cliff into the ocean – nature's convenient parable for the 1950s about the dangers of conformity and groupthink.

Although long debunked, 'lemming suicide' has reached such a pervasive place in pop culture that the Alaska Department of Fish and Game has a webpage devoted to dispelling the lemming myth. 'According to a 1983 investigation by Canadian Broadcasting Corporation producer Brian Vallee, the lemming scenes were faked,' Riley Woodford articulates on Alaska's Fish and Game webpage. 'The lemmings supposedly committing mass suicide by leaping into the ocean were actually thrown off a cliff by the Disney filmmakers. The epic 'lemming migration' was staged using careful editing, tight camera angles and a few dozen lemmings running on a snow covered lazy-Susan style turntable.'

But, as biologists point out and have for decades, this is simply not what lemmings do. The lemming scene has become infamous in its egregious and cruel use of animals to obtain footage of animals that wasn't even real to begin with.

More recent public backlash against the artifice of animal staging in documentaries came from the BBC's *Frozen Planet*. When the film debuted in 2011, viewers watched a pregnant polar bear traipse across the Arctic wilderness to find a den to give birth to her cubs. What viewers assumed to be the polar bear den in the Arctic – where the mother polar bear snuggled with her two-day-old polar bear cubs – was in fact a man-made den in a German animal park that was constructed to offer the best filmed footage of polar bear births and early den life possible. When viewers found out about the staged polar bear birth, they were outraged. This was the second time that the authenticity of Attenborough's work with the Arctic had come under fire – an earlier documentary from 1997 also showed a zoo-based polar bear birth that Attenborough and the filmmakers let audiences think was in a wilderness setting. The trust that the audience had in the authenticity of *Frozen Planet* was broken because audiences felt that they had been duped. Viewers thought they were seeing the real thing.

Many filmmakers argue that transparency in the process is the best way to invest audiences in the accuracy and authenticity of what they're watching. Simply telling audiences what is staged and what is not, and how certain footage was captured, invests audiences in the documentary. Jeffery Bosall has argued that on-screen disclaimers were the most ethical way to explain to audiences what has been staged and what has not. They would understand how the story was put together – and would not be expecting one thing and then be forced to grapple with another.

It's not so much that viewers object to polar bear cubs in a zoo – they object to thinking that they're seeing polar bear cubs in the pristine Arctic wilderness, then finding out that the scene was staged through investigative reporting, not through the filmmaker. And most twenty-first-century wildlife filmmakers would argue that this honesty – this transparency about the process – is what fundamentally constitutes best practices and invests audiences in a film's credibility and, especially, in its authenticity. Part of the attraction, then, of the wildlife cams is the transparency and honesty of what audiences assume they're looking at.

★ ★ ★

In contrast to the messiness of scripting a documentary about walruses, Point Lay's beaches provide a natural, easy viewing spectacle for internet enthusiasts, and audiences count down the days between migrations until their blubbery pinnipeds haul themselves out of the ocean and on to the Alaskan beaches for summer snoozing. But this still leaves open the question of how viewers make sense of the animals they're watching – animals that are far outside their everyday lives and experiences. For walruses, this question has been asked and answered, and asked again, for more than a millennium.

Historically, hauling out was how European hunters encountered these walruses, these creatures so curious and mythical and unknown in European circles. It ought to go without saying that just because the walrus was unknown to Europeans, this doesn't mean that the animals were unknown to indigenous populations in the Arctic regions. For at least 6,000 years, indigenous peoples have hunted walruses – walrus meat provided something like 60–80 per cent of their everyday diet, walrus ivory and bones were used for tools and weapons, and walrus skins provided covers for boats and dwellings.

Not only did walruses feature strongly in myths and other stories, but names for places in indigenous Arctic geography culturally triangulated to walruses. Today, scientists work with indigenous hunters to better understand walrus behaviour, because indigenous hunters have millennia of knowledge about the animals and their behaviour. Originally known to Europeans as *rosmarus* and *walrusch*, depending on the Dutch or Scandinavian source – probably derived from the Old Norse *hvalross*, meaning hairy whale – the animal was mostly unknown to Europeans for centuries.

In 1555, Olaus Magnus, the Swedish writer and archbishop of Uppsala, published *Historia de Gentibus Septentrionalibus* (A History of the Northern Peoples) while he was living at the St Brigitta monastery in

Rome. The book was a patriotic written testament of Sweden's folklore and natural history, eventually translated into four languages, reprinted and abridged six times, and it quickly became the text that defined Scandinavia to the rest of Europe's intelligentsia for centuries. Of the many oddities and curiosities in Magnus's book (and there were many!), his description of the *morse* – a hairy, carnivorous, walrus-like creature reported to sleep on cliffs while hanging by its teeth – is one of the most spectacular.

'To the far north, on the coast of Norway, there lives a mighty creature, as big as an elephant, called the walrus or *morse*, perhaps so named for its sharp bite; for if it glimpses a man on the seashore and can catch him, it jumps on him swiftly, rends him with its teeth, and kills him in an instant,' Olaus Magnus described the creatures. 'Using their tusks, these animals clamber right up to the cliff-tops, as if they were going up a ladder, in order to crop the sweet, dew-moistened grass, and then roll back down into the sea again, unless, in the meantime, they have been overcome with a heavy drowsiness and fall asleep as they cling to the rocks.'

According to Olaus Magnus, and as shown in the woodcut illustration in *Historia de Gentibus Septentrionalibus*, walrus hunters would creep up behind the animals and attach ropes to their tails. From a safe distance, hunters would rain rocks down on to the creatures, until they flung themselves into the ocean. Hunters would strip them of their valuable skins and the creatures would die.

But Olaus Magnus had never seen this hunting practice himself and, more to the point, he had never seen a real *morse*. (He was from inland southern Sweden himself, and never travelled to places where walruses actually lived.) More to the point, no one in the intelligentsia of sixteenth-century Europe had seen a walrus in its natural habitat either, so any zoological description or entry in a catalogue of natural history wouldn't depend on the author's own observations – the empirical *autopsia* so critical to later natural history. Rather, descriptions of the walrus would teeter somewhere between folklore and myth, as European intellectual circles became more and more convinced of the curious strangeness of the animals that inhabited the icy environs at the edge of their known world.

One of the earliest mentions of walruses, historian of science Natalie Lawrence notes, comes in 1250, when the German Catholic bishop Albertus Magnus described them in his *De Animalibus* as 'hairy whales' with 'the longest tusks' that they use to 'hang from the rocks of cliffs when they sleep'. The then-contemporary Polish diplomat and scholar Maciej Miechowita described these animals as being able to climb cliffs with their long teeth. The famous sixteenth-century Swiss naturalist Conrad Gessner, when not writing about fossils, included a description of a walrus in his *Historiae Animalium* (vol. IV), published in 1558. Although Gessner was sceptical about the sensationalised tales published by earlier authors, his walrus description was pulled together from a plethora of secondary sources including extrapolations of walruses based on a stuffed head on display in Strasbourg town hall, rather than his own experiences. Although some hunters published first-hand accounts of their encounters with walruses over the centuries, few made it into scholarly circles and fewer still were believed.

There's a bit of irony here in the way that Europe used walruses, but never knew them as an animal. For centuries, Nordic peoples used walrus tusks for chess pieces and shield decorations. Other walrus bits showed up in European apothecaries, and some were used as inlay for combs and brushes. Blubber was burned in lamps and made into soap, while tusks were sold off as unicorn horns and used as a traditional cure-all for poison. Only the indigenous Arctic hunters knew walruses before the animals were translated into dismembered parts for European commodities. The natural history of the walrus – what readers thought was real – hinged on the history that went into walrus myth-making.

Mainland Europe saw its first real live walrus in 1612, when a walrus pup was brought to Amsterdam, along with the stuffed skin of its mother. Dr Everhard Vorstius, of Leiden University, described how the pup 'raged like a boar', was soothed when placed in a barrel of water and ate mashed oats. (He then described walrus fat as tasting rather 'toothsome.') Confronted with such a curiosity and wonder of the natural world, it would appear that Renaissance Europe poked, prodded, then consumed the thing that it didn't comprehend, in what is perhaps the least subtle metaphor imaginable.

When it comes to the filming at Walrus Island Sanctuary, audiences who tune in to Walrus Cam are left with a bit of a conundrum perhaps not all that different from the question posed by Olaus Magnus, Conrad Gessner and Renaissance Europe. Most audience members have never seen a real live walrus before, and fewer still have seen it in its natural habitat. While viewers use their own powers of observation to take in the walrus-filled haulouts, there is still a bit of a gap in knowing just what those walruses are doing. Audiences fill in the missing story – perhaps it's similar to the walrus myth-making from centuries past – then project what sort of story they think they're seeing. The Real Walruses of Walrus Cam still need a guiding narrative to translate their real-ness into an authentic bit of nature and wilderness.

Wilderness, then, is nominally what Walrus Cam has set out to stream to its internet audiences, and what wildlife films from the BBC to Walt Disney have wanted to bring to viewers. Most twenty-first-century audiences watch scripted and live-streamed nature because such untouched, unexplored wilderness is exotic and inaccessible in their everyday lives. The desire to connect with the idea of nature underscores the very cachet that drives the popularity of wildlife films, and certainly wildlife camera feeds like Bear Cam and Walrus Cam. 'Longing for the authentic, nostalgic for an innocent past, we are drawn to the spectacle of wildlife untainted by human intervention and will,' historian of science Gregg Mitman suggests in *Reel Nature*. 'Yet, we cannot observe this world of nature without such intervention. The camera lens must impose itself, select its subject, and frame its vision.' Consequently, for people to look at nature (or Nature, with a capital letter) – whether through their own outdoor activities or through the world of documentaries – is a series of trade-offs is involved about the authenticity of the nature or wilderness under debate. 'Cultural values, technology, and nature itself have supplied the raw materials from which wilderness as artifact has been forged,' Mitman argues.

So where does that leave us in the ways that we watch wilderness and nature and whether we think such watching is authentically real?

If the point of nature documentaries is to introduce audiences to well, nature, in a more scripted, curated form than Walrus Cam, what are the acceptable ways of doing so? Add to that the business – the commodification of animals and natural wilderness, that is – and the issue of parsing what is genuinely fake nature is no small feat. The end goal of both seems to be the same – but the means and methods of authentic viewing are incredibly different.

Back on the Round Island beach, the walruses don't seem to care whether audiences watch them *Frozen Planet*-style or through Explore.org's Walrus Cam; the migrating

CHAPTER SIX

The Great Blue Whale

Call her Big Blue. She's just that. She's 26m (85ft) of big blue whale living at the Beaty Biodiversity Museum in Vancouver, Canada.

Size is the most defining attribute for blue whales. Big Blue and her fellow *Balaenoptera musculus* are the largest organisms to have lived on Earth. They are bigger, even, than the massive Mesozoic sauropod plant-eating dinosaurs like *Argentinosaurus*. *National Geographic* and the Royal Ontario Museum put the size for most blue whales at about 30m (100ft) in length, with them eating something like 3,600kg (7,900lb) of krill a day. Every time a blue whale opens up to take in a mouthful of krill, it is the largest single feeding event that has ever happened. Blue whale calves gestate for a year, are born weighing three tons and gain 90kg (200lb) a day for the first year of their lives. This organism is so enormous that it's hard to internalise just what these impressive stats really mean.

But Big Blue, such as she is, is more than just her measurements and factoids. By necessity, she and others of her species are built out of superlatives, similes and parallels. ('Blue whales are longer than two trolley buses parked behind each other.' 'Their hearts are the size of a car and the arteries connected to the car-sized heart are large enough that a human baby could crawl through them.' 'A blue whale's tongue weighs as much as an elephant.') Because Big Blue's magnitude is so out of humans' life experience, these analogies are attempts to make her size accessible to us – a species that would need something like 15 of its individuals standing on each other's shoulders to even look Big Blue in the eyes.

Blue whales are found across the globe in the Atlantic, Indian and Pacific Oceans. At the beginning of the twentieth century these colossal cetaceans were abundant, and biologists estimate that whale populations in the mid-1800s were somewhere between 250,000 and 350,000 individuals. However, decades of unabated, industrialised whaling in the late nineteenth and early twentieth centuries decimated blue whale numbers, as they did a plethora of other whale species, as whalers made their fortunes by hunting the animals for oil and blubber. Between 1904 and 1967, a total of 350,000 blue whales were hunted from the Antarctic Ocean. In the 1931 season alone, whaling ships caught something like 29,400 blue whales in the Antarctic.

In the 50 years between 1920 and 1970, scientists estimate that on average, blue whale populations lost 20 per cent of their body mass, essentially becoming smaller because of commercial whaling. Under pressure from conservationists and scientists horrified by whaling's effects on marine populations, 1972 saw blue whales achieve legal and conservation protections to save them from being systematically hunted. Even under such protections, however, current blue whale populations have been slow to rebound, with only 1 per cent of the pre-whaling population alive today. For most of their recorded history, whales have been defined through their relationship to whalers and the whaling industry as, for better or worse, most scientific knowledge about whales has come from the slaughtered animals.

In the twenty-first century, blue whales are no longer a mere commodity of the whaling industry, valued for their meat, blubber and oil. In the post-whaling world of the late twentieth and early twenty-first centuries, blue whales have become an icon of conservation efforts – the charismatic megafauna of the ocean. Where savannahs have elephants and mountain forests have pandas or gorillas, the blue whale is the ocean's cultural ambassador, reminding *Homo sapiens* that it holds blue whales' destiny in the hands of its environmental policies and practices, and have for over a century.

★ ★ ★

Big Blue died just off the coast of Prince Edward Island (PEI) in 1987, stranded on the Canadian beach near the village of Tignish. When the stench of 80 tons of rotting whale flesh assailed the nostrils of island locals, the quick consensus was to bury the corpse with backhoes and hope that the reek disappeared into the red, sandy clay of PEI's beaches. It took two days and serious earth-moving equipment to cover up the animal. For two decades, the whale carcass remained in its shallow grave – unmarked, unmapped and generally forgotten.

In 2007, however, whale biologist Dr Andrew Trites of the University of British Columbia (UBC) decided to excavate the blue whale skeleton to be the centrepiece of the newly opening Beaty Biodiversity Museum on UBC's campus. In media interviews, Dr Trites described his pursuit of a blue whale skeleton as the fulfilment of his life-long dream of finding, cleaning and curating such an exhibit. Exhibits of blue whales raise the prestige and scientific cachet of museums and never fail to impress visitors.

In order for the Beaty exhibit to work, Trites argued, it had to be a real skeleton of a real blue whale. None of the business of creating life-size blue whale models like those that had hung in the hallowed halls of the American Museum of Natural History and the Smithsonian for decades. Those were fine for the education of museum visitors, but having a real skeleton would do that *and* it would be authentic to boot. Only the Real Thing – as he saw her – would do.

Worldwide, to date there are only 21 blue whale skeletons displayed as part of museum exhibits. Blue whale skeletons hang, trophy-like, in natural history museums from the Marine Discovery Center in Santa Cruz, California, to the South African Museum in Cape Town. The Canterbury Museum in Christchurch, New Zealand, had a blue whale skeleton in its Garden Court from 1976 to 1994, before it was moved to storage pending its eventual return to a newly refurbished museum hall. The Natural History Museum in London has one of the oldest skeletons of a blue whale, collected from Wexford Bay, south-east Ireland, in 1891 and put on display in 1934. (The whale, christened Hope in the twenty-first century, was moved to the Hintze Hall in 2017 as part of the museum's recommitment to exhibiting specimens that show off the beauty and wonder of the natural world.) The Royal Ontario Museum in Toronto collected two blue whales, which had stranded in 2014, from a Newfoundland beach.

It's a rare thing for a blue whale to strand itself in circumstances that are conducive to it being collected by a museum – and even rarer for a museum to have the resources and wherewithal to collect, clean and curate its skeleton. Having a real blue whale skeleton on display would certainly establish the newly built Beaty Museum's scientific legitimacy and authority. 'In British Columbia, there's probably fewer than 10 that still live off of the shores of BC,' Trites pointed out in the 2011 documentary *Raising Big Blue*, highlighting how rare the animals are. 'In eastern Canada, it's in the low hundreds. It's such a rare, rare animal. And the opportunity to find one when it does die is even rarer.'

In *Raising Big Blue*, Trites scours several locations in the hopes of finding his Blue, but 'it's not until he caught a whiff of the old story' that the PEI whale became a contender. In mid–December 2007, the Beaty Museum sent a recon team of four scientists to PEI to find the buried skeleton and to determine if recovery was possible. Six months later, in May 2008, Trites, his team and his intrepid volunteers dug in and exhumed the remains of the buried PEI cetacean. Trites thought the entire project might take two weeks: go to PEI, find the whale bones, dig them up and ship them back to Vancouver – easy, peasy.

However, the project of removing the whale from the beach immediately became much more involved and difficult than Trites had originally imagined. The clay-infused sand had not been conducive to the whale's decomposition, as such soils are low in oxygen and hence prevented bacteria from growing and breaking down the whale. (Trites had hoped that enough time had passed from when the whale had been buried for the fleshy parts to have decomposed, so that he'd be left with a clean, articulated skeleton to pluck from the sand and hang in the museum.) To add a bit of insult to injury, instead of excavating just whale bones, team members found that that they were excavating exactly what the PEI locals had originally objected to – the smell of still-rotting whale carcass.

'I tend to think about whale decay as a process from sashimi-grade fresh to burn-your-clothes disgusting. At the outset, there is something about the volatile organic compounds, like wax esters or hydrocarbons, locked in blubber, that appeal to me – a fresh smell associated in my mind with dissection and hands–on discovery,' Smithsonian curator of fossil marine mammals Nick Pyenson said when I asked him to describe the stench of dead whales, 'and that tends to be a strong smell associated with whale bone collections because the lipids generally are pretty difficult to entirely pull out of whale bones.'

Working in raincoats and galoshes, the team found itself working with close to 80 tons of dead-but-not-gone putrid mammal. After two weeks of efforts to extricate the carcass from the sandy clay, Trites organised for the entire set of remains of the blue whale to be transported back to the Beaty in refrigerated trucks – to stall the decay now that the whale was exposed to air and environmental conditions again – to Vancouver museum was located more than 6,000km (3,730mi) away from PEI.

Once the whale bits arrived in Vancouver, the flesh, muscle, and oil had to be stripped from the bones before the skeleton could be hung in the Beaty. (In an odd bit of art imitating life, the project of excavating and mounting the PEI whale quickly became its own set of superlatives – raising 'Earth's Largest Animal' from 'Nature's Biggest Stink' in a project hyped as an 'Impossible Cleanup'. The entire enterprise could be most succinctly described as 'Herculean'.) Cleaning the skeleton wasn't easy, but it was rather straightforward. Degreasing her was something else entirely.

To understand why cleaning whale skeletons is so difficult and is such an involved process, it's important to keep in mind that whale bones are extremely porous – much more so than other mammalian bones. When whales are alive, their spongy bones are filled with oil. However, not all stranded whales end up with such greasy skeletons. The Prince Edward Island blue whale was particularly problematic because of the type of soil that she had stranded herself on. (Some whales are stranded on soils more conducive to natural decomposition practices, and curators and scientists can simply bypass these steps entirely.) For 20 years, this blue whale's oil had been going rancid in PEI's clay sands, and when Mike deRoos, the blue whale Project's Master Articulator, popped open the cargo containers in Vancouver, the reek was apocalyptic.

DeRoos and his team began cleaning the bones using special degreasing tanks, spraying down everything with a degreasing enzyme to try and break down the oil. 'We used industrial cleaning and de-greasing enzymes solutions produced by a company called Novozymes,' deRoos explained to me over email. 'These were primarily lipases mixed with detergents that are typically usually used to clean restaurant floors and grease traps and to clean up oil saturated sites *in situ* among other things. We also used bacteria designed to further break down and digest the products of the enzyme reactions to water soluble waste products that could then be flushed from the bones. It took a lot of attempts to refine the process.'

But the bones still seeped. They sprayed down the bones with the enzyme. But the bones still seeped. The team built a 2,500gal (11,356-litre) bath filled with marine bacteria that could digest oil. Even then, some of the bones still seeped. After months of struggling to keep the bones from becoming a pulverised, soupy mess in those baths, and up against the deadline of the museum's opening night, the whale-cleaning group consulted a team from the university's microbiology department in November 2008 – the microbiologists recommended that the cleaning crew turn up the heat in the baths. And yet some of the bones still continued to seep.

Eventually, deRoos put the larger bones that were veritable oil reservoirs into a vapour-degreasing process that seemed to do the trick. The team had lost some of the whale bones over the course of the cleaning as they had simply broken down and melted away with the intense enzyme treatment. Still determined to see this specific blue whale skeleton hang in the Beaty, Trites had duplicates of destroyed bones, like the skull, sculpted in plaster. (Creating a replica replacement of the skull required 60 moulds of different parts of the skull to be cast, then put together.)

'Once the skull is painted,' the *Raising Big Blue* documentary points out, 'no one will be able to tell the replica from the real.'

From January through April 2010, deRoos began to put the bones in order, and to lay out how the whale would be articulated inside the museum. Between April and May 2010, the team installed the skeleton, with the final logistical nightmare being a game of high-stakes Jenga to get the whale bones through the front door and into the museum. The bones were hoisted into place and, at 26m (85ft) in length, this is the largest blue whale on display in Canada. Captain Ahab might have chased his white whale 'round Good Hope and Perdition's flames, but at least he didn't have to deal with the headaches of whale rot, enzyme baths, liquefying bones and museum deadlines that had beset Trites and his team in their quest to put that PEI whale in the Beaty. Finally, on opening night, as the whale skeleton hung in all its glory from the Beaty's ceiling, Trites could take a bow as a result of the success of his project. The whale was christened Big Blue.

Today, all 26m (85ft) of Big Blue hang from the ceiling of the Biodiversity Museum, comfortably reigning supreme over the Beaty's two million specimens. She is indeed, visitors are told, a real blue whale skeleton, with the documentary of her discovery and cleaning playing on loop to prove it. But, like so many things, pinning down just what makes Big Blue authentic is incredibly tricky. What makes Big Blue – and other museum exhibits – the Real Thing is the commitment to scientific and cultural authenticity that underlies them, and has for more than a century.

★ ★ ★

The desire to show audiences real whales and give them real whale stories was – is – certainly obvious in museums like the Beaty here in the twenty-first century. But showing off real whales has a long history that reaches back to circus and sideshow acts that peddled whale shows along North America's highways and byways in the late nineteenth and early twentieth centuries. As early as 1873, the Great International Menagerie, Aquarium, and Circus, for example, touted its exhibit of 'A Leviathan Whale, a grand and magnificent specimen, the King of the Deep' as '… the only show in the world that exhibits a WHALE'. And by 'WHALE', the Menagerie meant an actual dead whale that was quasi-taxidermied and carted around from one American town to the next. Not to be outdone, the Burr Robbins Circus exhibited a giant papier-mâché whale a few years later. The two whales duelled their way across the United States for almost a decade, competing with each other for audiences and often picking up on each other's publicity.

At each town's stop, the Great International Menagerie would build a staging area for the whale exposition, with local carpenters as the exhibit's harbingers. (It would appear that the Great International Menagerie, Aquarium and Circus was an early adaptor of the 'If you build it, they will come' business strategy.) The show, the bustle of activity and the anticipation of the WHALE as something so singularly extraordinary never failed to draw huge crowds. The draw was enormous and, more to the point, the experience was personal. Step right up, ladies and gentlemen, and see for yourself this creature of the deep. You, your very own self, can tell others what real whales look like because you saw them here. The Menagerie's King of the Deep arrived in Columbus, Ohio, on 8 March 1881. The headlines in the *Columbus Dispatch* screamed, 'He is coming! He is coming! The Monster Whale! The Monarch Supreme of the Ocean! The Giant of the Gigantic Creation of the Universe! Don't Fail To Bring The Children!'

When the Menagerie's whale showed up, there was a mob scene. ('It requires a great deal of elbow room, because his whaleship is sixty feet long in the clear,' the *Dispatch* reported.) The exhibit focused its efforts on securing and displaying the whale, but not so much on explaining anything about whale biology to its audiences. Interestingly, though, it was the story of how the whale went from a living marine mammal to a dead exhibit travelling through Columbus that really captivated nineteenth-century menagerie-goers.

Visitors invariably wanted to know how the exhibit was possible. They wanted to know exactly how the whale's entrails were removed and replaced with first ice, then chemicals, how the sawdust underneath the whale corpse kept the leviathan's leaking in check, how yet more chemicals on the outside of the whale kept the skin from decomposing and how iron hoops within the whale's body kept His Whaleship from collapsing. The *Columbus Dispatch* reassured readers that the exhibit was 'free from unpleasant odor'.

In short, the whale wasn't really so much about the whale, but about how the whale's keepers and revenue collectors had so effectively cheated the decay of death, thus introducing audiences to the whale as the Real Thing, of sorts. The Menagerie whale was open for business at each railway stop, from 9 a.m until 9 p.m.

Although the more 'respectable' display of whales in a scientific context, for more 'cultured' audiences, began to enter museums like the American Museum of Natural History at the turn of the twentieth century, the era of *Music Man*-like whale showmanship was far from over. In the 1920s, Hugh Fowzer – an experienced, ever-enthusiastic sideshow entrepreneur – hatched the idea of putting a dead stuffed whale on a railway carriage, certain that audiences would leap at the chance to see such a rare oceanic creature.

(One Carl Terrell, a motordrome rider, claimed to have given Fowzer the idea for a whale show based on one Terrell claimed to have seen in Omaha in 1921 on a converted showboat. Terrell's claims were never taken very seriously.) Those earlier touring whales from the Great International Menagerie had been so successful, Fowzer reasoned, that adding one to his own enterprise ought to be practically money in the bank.

Fowzer teamed up with Wingy Counts, from Venice, California – known as 'One-Arm Wingy' in the papers – a sideshow fixer who was familiar with the ins and outs of putting together a marine exhibit. He had established several marine exhibits for other owners along the West Coast and his credentials, such as they were, were in perfect order. (Fun factoid: Counts gained his showmanship acclaim by wrestling an octopus in live shows.) Somewhere between 1921 and 1922, Fowzer put Counts in charge of pulling together his whale exhibit, and somehow Counts managed to secure a somewhat embalmed whale. Fowzer and Counts took their whale on tour, and within a couple of years the enterprise was so successful that they had two travelling exhibits that crisscrossed American's railways.

As Fowzer and Counts were working on their menagerie act, a second set of sideshow entrepreneurs was working to start taking its own dead whales on tour. But this begs the question of just how, exactly, one preserves a whale, which brings us to the story of Mr M. C. Hutton and wax-museum operator and self-made entrepreneur Mr Harold L. Anfenger. The estimated cost of embalming a whale in 1928 was staggering – something like $10,000 – and Hutton balked at ponying up this kind of capital for what could most charitably be described as a whale of a gamble. Entrepreneur that he was, however, he offered four investors the 'opportunity' to buy into the whale exhibit for a mere $2,500 apiece. Money thus secured, Hutton bought a whale from a whaling ship's successful hunt and immediately had the cetacean towed into harbour. The commander of the whaling ship, Captain Dietrich, demanded – and more to the point collected – $1,500 upfront for delivery of said whale and had nothing else to do with the entire business.

This is where the story of this particular whale gets a little bit complicated. Before his dealings with traveling whales, Anfenger had established a rather lucrative whale-based business in Long Beach, California. Anfenger contracted a crew of boatmen to anchor dead whales close to the beach, and for days the crew would row beachgoers out to see the carcasses at 25 cents per person. Tethered just offshore and subjected to the elements, the whales began very rapidly to shuffle off their mortal coils and the stench was considerable; consequently, Long Beach residents were relieved to have the whales towed back out to sea at the end of Anfenger's enterprise. When Hutton and Anfenger's whale was delivered, they argued about how to best preserve it for exhibition, eventually hiring an embalmer who poured barrels of formaldehyde and salt into the cetacean, claiming it was a sure-fire preservation method. The whale promptly exploded.

Undeterred, Hutton and Anfenger doubled down on their efforts to find ways to round out their travelling sideshow with a real dead whale, convinced that audiences would queue to see something so mysterious. And in fact, they eventually assembled a rather formidable whale pod of more or less preserved cetaceans, once the chief embalmer, Mr Griffith, had established the best ratio of formaldehyde and salt, thus ensuring that the whales wouldn't decay or detonate. Still, exhibiting dead whales to a curious public was a lot of work and required constant attention; one of the everyday chores for exhibit workers was to administer daily doses of formaldehyde and salt, shooting the solution into the fleshy bits of the whale with foot-long hypodermic syringes. 'The giant tongues were the worst source of foul smell,' circus historian Fred Pfening Jr. notes in his history of circus menageries, 'but the tongues were deemed too important to remove.'

Fowzer's foray into these nautical show enterprises was a raging success – his gamble more than paid off, as did Hutton and Anfenger's. Fowzer had two whale units and the Pacific Whaling Company had nine. Like their nineteenth-century counterparts, the two whale exhibits often capitalised on each other's publicity. The exhibits travelled throughout the United States during the first half of the twentieth century with incredible financial success. The first of the Pacific Whaling Company's units grossed over $100,000 in income during the six months of its initial exhibit in 1928. Eventually, Pacific bought out Counts to put a halt to the sniping of whale publicity, and by the 1930s, the Hutton and Anfenger enterprise diversified to support several different travelling units operating throughout the United States. Former sideshow lecturers were dressed up in nautically themed sea captains' costumes and schooled by a circus performer, Arthur Hoffman, in appropriately salty ocean vocabulary.

As the success of the whale exhibits gained momentum, the entire enterprise began to expand in scope. One of the new units was called a Modern Noah's Ark, and it boasted a two-headed cow and a female mentalist 'who will guess your age'. Another show in 1937 advertised 'Mammoth Marine Hippodrome and Congress of Unbelievable Biological Exhibitions' – its headlines boasted a mermaid, a flea circus, mummies from Egypt and a unicorn. Amid such specimens, something as 'mundane' as a real, stuffed and sort-of leaking whale didn't quite headline the acts any

more. Audiences no longer looked at whales in these shows
with awe and wonder, as they had at His Whaleship in the
1880s. By the end of the 1930s, Hutton and Anfenger
began to cut back on their exhibits and by the 1940s, the
legacy of whale showmanship was all but over. What
counted as spectacle, it would seem, had changed. When
Anfenger sold the railway carriages in the 1940s, disposing
of the whales inside them was a serious problem for the
new owners.

It's undeniable that, between the Great International
Menagerie, the Modern Noah's Ark and Fowzer's
travelling enterprises, thousands of people saw real
whales. But they saw them the way they would see
anything else at a travelling sideshow. These sideshow
whales did not have their own unique biology and
history, and were objects stripped from their natural
contexts – they were merely the means to financial gain
and simply spectacle for spectacle's sake. Where whaling
had commercialised the different parts of whales, like
their oil and bones, the early twentieth-century sideshow
had monetised the possibility of seeing something real,
such that it was.

These whales were real, sure. But they weren't the Real
Thing – they didn't command or carry any sort of cultural
or scientific cachet other than that of a sideshow oddity.

★ ★ ★

Back in the museum world, there was a burgeoning
interest in introducing audiences to the life and history of
whales, especially blue whales. What drew audiences now
was authenticity – scientific authenticity, to be precise.
Unlike the freewheeling world of sideshows, museums
run by conservators, curators and biologists wanted
authentically modelled and scientifically credible

specimens – and none more so than the blue whale. Simply due to the size of animal, the awe and wonder of nature provided its own cetacean propaganda. Isn't nature amazing, museum whale exhibits suggested to their visitors, and isn't nature spectacular? It was the same awe and spectacle of nature that, a few decades later, would inspire Disney's *True-Life Adventures*. What made the exhibit authentic was the specimen accurately rendered, preferably in context.

Museums were obsessed with the idea that their whale models – be it plaster, fibreglass, skeletal or papier mâché – must be as close to the real whales as possible. Museum whales, sniffed the institutional staff, were certainly not going to be amateur, pumped full of formaldehyde and salt. These would not be sideshow specimens that leaked on visitors and dripped with the failed dreams of entrepreneurial avarice. Museum whales, they contended, would be exhibits that combined truthful, accurate and authoritative accounts of whale life and biology – organisms with the moral and aesthetic demands of accuracy and, most importantly, based on real cetaceans. 'This process of categorization – real or fake, flesh or *paper-mâché*, true or false – was persistently present for exhibitors,' historian of science Michael Rossi describes of these early whale models, 'not least of all when [museum curators] endeavored to construct honest replicas of tremendous creatures, which spent most of their time obscured beneath the ocean.'

Preserving a whale is both insanely hard and incredibly complicated because its biology is so different from that of other animals. In land animals, the outer skins – the fur, feathers, scales and shells – give the animals their surface realism. Stuffed, posed and displayed, land animals in dioramas became expected features for museum-goers in the early twentieth century. By seeing what these animals would have looked like in real life, audiences could easily believe that the still-life dioramas in museums were, in truth, authentic three-dimensional still-life portraits of the natural world. 'The animal is dead but not gone, refashioned but fundamentally still available,' historian Rachael Poliquin suggests in her history of taxidermy, *The Breathless Zoo*. 'The various genres of taxidermy were all created to satisfy a variety of longings for continued connection. The desire to capture beautiful forms, to tell stories about their importance, or to offer lessons in natural history all fundamentally shape how the resulting animal-things will be perceived and understood.'

Unlike other mammals, whales are completely devoid of hair (except for on their snouts, when they're born), and unlike other mammals, whales shed their skin by constantly sloughing off large sheets of it. Consequently, while the furry or scaly skins of most animals are relatively easy to preserve through taxidermy, the whale's dermis simply can't be preserved, due to its biology. (Sideshow organisers found this out the hard way, with their constant battle to keep the whales from leaking and the skin from decaying.) If a museum wanted an authentic whale that looked life-like, the museum would have to settle for a display that didn't necessarily contain parts of real whales.

This is where museums bifurcate both in what sort of real blue whale they show visitors and in how they show it. Some museums have simply opted to display a blue whale skeleton. This offers visitors the opportunity to see the animal's real bones – to gape at its gargantuan size and to wonder at its impressive shape. ('And it smelt. How it smelt! It smelt as if forty thousand freezing and soap works were holding a reception, with sewer systems as guests,' Canterbury Museum curator Edgar Waite wrote in 1908, while he worked to clean the stranded blue whale that would eventually go to the Christchurch Museum in New Zealand.) With blue whale skeletons in their collections, museums could also appeal to scientific researchers who would measure and study the bones – such was the case with the British Museum in the early part of the twentieth century, and the Beaty Museum in the twenty-first.

The feedback between scientists actively 'doing' scientific research and the display of the blue whale as a scientific object ensured that the museum – as an institution – maintained its scientific authority and credibility. (As well as the outreach value of a whale seen by public audiences.) In other words, museum visitors would have confidence in the real-ness of the whale skeleton on display in the museum – and the prep that is implicit in displaying the skeleton – in a way that they wouldn't when having themselves a gander at the travelling whales in railway exhibits.

But whale skeletons – even real ones – still require visitors to imagine the skeleton with muscle, skin and flesh. The bones were real ones, sure, but a skeleton suspended from the ceiling is not how one would really see a blue whale in nature. For the American Museum of Natural History as well as the Smithsonian, at the turn of the twentieth century the answer to this conundrum was to build replica models of whales – and build models they did. Indeed, building a replica of a blue whale – twice – introduced a whole host of issues to American Museum of Natural History curators. The original blue whale that was finished and displayed in the Museum in 1907, as well as its 1969 counterpart, were tricky to make real because a replica would not only have to look like a blue whale, but to behave like one as well.

★ ★ ★

'They are building a whale at the Museum of Natural History,' the *Wilkes-Barre Record* of Pennsylvania reported on 4 January 1907, 'from wooden strips, iron rods, piano wires, paper, and glue. They are carpenters, horsesmiths, and wallpaper hangers and when the work has been done in rough, the naturalists will give the finishing touches.' This American Museum of Natural History blue whale model was built out of expert science and natural history, as well as iron, wood and canvas, and a very believable papier-mâché exoskeleton of sorts. 'First the whale was papered … he got a nice coat of heavy pulp sheets, then a thin layer of red fibre with a little manila in it and, last of all, a thin coat,' the paper informed its readers. 'This was one of the queerest bits of paperhanging ever done in New York.'

The curators who commissioned this original blue whale, as well as the engineers and artists who built it, depended very much on the notes, photographs, plaster casts and measurements that whalers, scientists and naturalists had collected in the field. This was the case for all whale exhibits, as whales weren't photographed underwater in their natural environments until as recently as the 1970s. Since curators were not able to transport whale corpses to study (the way that curators of other, smaller species might be able to), they relied heavily on notes, photographs and stories from both scientists who worked with whales and whalers who had a lot of practical whale knowledge. When the American Museum of Natural History built its model blue whale, it relied on a combination of scientific and artistic expertise to be properly authentic.

But, again, since the blue whale is such a huge animal, measuring, photographing and observing the mammal was no small feat. Blue whales are hard to measure and hard to weigh. It's tricky to capture the entire whale in a

THE GREAT BLUE WHALE

single photograph, and in a world of black-and-white photography, capturing the whale's blue-grey hues was impossible. As most whale observations were made by scientists on whaling expeditions (some scientists were even posted to such expeditions because this was one of the few ways to observe whales while they were alive), there was very little opportunity for scientists to observe whale behaviour. Most of the whales that were studied were the pre-processed, hunted carcasses amassed by whaling teams. Taken together, these logistical issues simply made it difficult for scientists to know things about whales, and if it's difficult to know things, it's even more difficult to accurately translate that perfunctory whale biology into a blue whale exhibit.

'In producing their model, exhibitors at the American Museum employed a patchwork of overlapping … techniques,' Michael Rossi points out, 'to argue that their fabrication was as authentic – as truthful, accurate, authoritative, and morally and aesthetically worthy of display – as an exhibit containing a real, preserved cetacean.' The resulting blue whale replica was lauded by *Scientific American* and *Outlook* magazines for its authenticity and commitment to detail. It inspired generations of New York City schoolchildren and offered visitors a glimpse into life in the world's deep, deep oceans.

Fifty years later, the museum opted to update the blue whale by replacing it with a new model. On the one hand, the new model would reflect the scientific updates of five decades of research. On the other, the Smithsonian had recently installed a (28m) 92ft blue whale model and the American Museum of Natural History wasn't about to be outdone. (The 1907 model 'only' measured 25m/82ft in length and certainly wasn't going to be 'one-upped' by the Smithsonian.) All of the questions about real-ness and believability of the whale that fronted the museum administration in its original blue whale model were asked, answered and asked again in its mid-century update.

When curator Richard Van Gelder of the American Museum of Natural History described his experiences of working to build the model of a blue whale for the museum (it went on display in 1969), he was frank about the parameters and limits of what it was possible to know about blue whales and certainly what it was possible to accurately build in a museum. ('So far as accuracy was concerned, I couldn't see much wrong with it, mainly because I had never seen a blue whale.') But despite the limits of blue whale knowledge, museums didn't just make up blue whales and stick them in the main halls of their buildings. The limits of whale knowledge resulted in accurate – authentic – blue whale exhibits being under constant negotiation.

For the American Museum of Natural History in the 1950s, trying to figure out how to modernise its blue whale exhibit, firstly there was the issue about the sort of pose to put the blue whale in. Should it be swimming? Eating? Diving? For years, the museum administration and its curators argued back and forth about the best way to balance the demands of authenticity and cost. In a burst of cynical disgust and snark, Van Gelder suggested to his bosses that the best way – the most *authentic* way – to display the blue whale would be to have a carcass of one on a sandy beach, with the sounds of gulls and sandpipers twittering while they picked away at its flesh. This was the way that most people had seen the elusive blue whale in nature – hence it would be truly real. To hear Van Gelder describe it, this was his modest proposal. By offering up something so inherently disgusting and ridiculous, everyone could go back to the business of creating an exhibit that was sensible.

But Van Gelder underestimated his bosses' stinginess. Because it was an exceptionally cheap way to display a model, the 'higher-ups', as Van Gelder called his bosses, green-lit the dead whale carcass exhibit with unabashed enthusiasm. Van Gelder spent months trying to convince the museum administration that what he had suggested in jest was in fact a horrible idea. He finally concocted a way to undermine the project by pitching the circle-of-life carcass idea to the Women's Committee of the American Museum of Natural History, a group that exerted considerable influence on the museum's administration due to its incredibly successful fundraising efforts. These ladies could make or break an exhibit.

'Finally, I told them about our "wonderful" beached-whale exhibit. I waxed poetic with word pictures of the beast. I told how the cries of the sea birds would slowly die out as sunset approached and then the ghostly glow of the bacteria would take over until at dawn, once more, the crash of the waves, and the rising chorus of hungry gulls would again take the shore,' Van Gelder recalled in 1970, several years after his presentation of the dead whale exhibit to the fundraising luncheon for the Women's Committee. 'I dropped my voice to a conspiratorial whisper. "We are even planning something never done before. A gentle breeze will waft the odor of the sea toward the visitors, to complete the attack on all the senses, and we are even going to try to simulate the odor of the decomposing whale, so that all can share in this wonderful experience *in totality*."'

According to Van Gelder's account, many of the ladies came close to losing their Chicken-à-la-King at this prospect.

As Van Gelder had hoped, there was only so much authenticity that museum-goers (and museum fundraisers) would tolerate. His boss was assaulted with a slew of complaints about an exhibit that showed a whale corpse. 'Why,' the committee wanted to know, 'did we have to have a beached whale – a simulated dead one? Why couldn't we have a whale that looked like a live one (and wouldn't smell …)?' Thus, the dead whale exhibit died its second death, reopening the question of what sort of blue whale model the American Museum of Natural History should have and what it should look like.

At the end of the day, The Whale, as the American Museum of Natural History model was dubbed, managed to balance authenticity, cost and administrative demands when it was posed in a jack-knife dive, attached to the museum's ceiling with a stout pole. The scientific authenticity imbued in the model seems to have endured well the decades of blue whale research. The Whale was a public hit – 35,000 people visited the first Saturday that the exhibit was open to the public.

★ ★ ★

The success of The Whale followed a precedent that had been set five decades earlier, when the American Museum of Natural History displayed its first blue whale replica in 1907 – and it turns out that 1907 was a very busy year in the world of artificial whales. While the American Museum of Natural History was taking bows in the scientific and museum worlds for its first life-size model of a blue whale, the original 25m model – the one constructed from wood and angle iron, covered with papier mâché – New York engineer William Muhlig initiated a lawsuit in Manhattan's municipal court over the 'Only Preserved Greenland Whale Ever Seen in Captivity'.

Muhlig's whale (its species is never specified) was a specimen that had, as reported by *The New York Times*, been captured off the shores of Greenland and stuffed in Hamburg, Germany – all at great expense. The two owners of the whale – Christopher Rebhan and August Brahn – had offered Muhlig a share in the taxidermied cetacean. For only $1,500, Muhlig could buy the opportunity to become the whale's manager as it toured the United States, collecting a salary of $20 a week and a quarter of the profits from the show in return. Eager to make such a lucrative deal, Muhlig fronted $500 and the whale began its tour of the eastern United States. Things with the whale went along swimmingly, as it were, until Harrisburg, Pennsylvania, when Muhlig discovered that the whale was not in fact a preserved Greenland whale at all. It was, according to court documents, 'a wooden dummy covered with canvas'.

Enraged, Muhlig put the fake whale up for auction, netting a paltry $99 for his efforts. (Mrs Brahn, August's wife and Muhlig's cashier, promptly seized the cash as her salary.) In an effort to recoup his originally invested $500 from Rebhan, Muhlig initiated legal action in the Manhattan municipal courts. Not only did he lose his lawsuit, but the courts also had a field day mocking him for his gullibility in falling for the scheme in the first place. Muhlig self-righteously pursued his claim, and on 7 June 1907, the Appellate Term of New York's Supreme Court ruled that the lower court had subjected Muhlig to undue ridicule in the case's original ruling.

However, this begs the question of how Muhlig's fake whale was so very different from the then-newly installed faux cetacean at the American Museum of Natural History. 'How can it be that [the AMNH's] wooden whale was scientifically authentic and admirable, while Muhlig's wooden whale was silly, and even actionable?' Michael Rossi ponders, neatly illustrating the dissonance of the two whales. 'Visitors to the American Museum were actively encouraged to pretend that ... whale was the real thing. Muhlig's whale was humbug; [the whale at the AMNH] was cetology – but how?'

What makes one of these fake whales more real than the other? The intent? The methods of making it? The materials? The story that visitors take away from their experience? Some combination of everything?

How we think about the question of fakery or authenticity is important. Simply asking whether something is 'real' or 'not' (even 'authentic' or 'not') is not really helpful, nor is it particularly insightful. Each of the genuinely fake whales – like those from the Great International Menagerie and the American Museum of Natural History – balanced a series of trade-offs in cost, audience expectations and realness, putting each of the whales at a different place along the continuum of authenticity. As whale curators and showmen have found, there's only so much authenticity about whales that audiences are willing to tolerate – no leaking, dripping or smelling – even if those things are just as 'real' as the other parts of an exhibit.

Likewise, there's an expectation that a whale in a museum will be scientifically accurate and have a dignity and gravitas about it that certainly wouldn't be found in a circus sideshow. Showing audiences the Real Thing means showing audiences that the Real Thing is always a work in progress. For over a hundred years, blue whale skeletons and models have afforded non-experts the opportunity to see and wonder about the largest animal that ever lived on Earth. From the flimflamming whale sideshows to the fibreglass replica at the American Museum of Natural History, seeing the Real Thing has invested audiences in whales in ways that wouldn't have been possible otherwise.

This brings us back to the story of Big Blue at the Beaty Museum of Biodiversity in Vancouver and Dr Andrew

Trites's Ahab-like quest to make sure that the Real Thing was what was on display. Even though Big Blue's skull is a plaster model and other parts of her have been repaired, fixed and artistically rendered, there is little to dispute that visitors are seeing a real blue whale skeleton. What really sells the real authenticity of the Beaty's blue whale is not just Big Blue's skeleton – it's also the story that goes with her.

CHAPTER SEVEN

And Now It's the Real Deal

The premise of *Antiques Roadshow* is simple. Debuting in 1977, the television programme had expert appraisers travel throughout Britain and assess the value of antiques that locals brought in. The appraisers authenticated any collectibles and offered a brief history of the craft and context of each piece. *Antiques Roadshow* was an exciting way of sorting out genuine treasures from hoarded bits and bobs gathering dust in British attics.

Currently in its fortieth season, with spin-offs in Canada and the United States, the show is a raging success. Most of the objects dragged into *Antiques Roadshow* are just what they appear to be – kitschy heirlooms with quirky family stories or antiques picked up as a bit of a lark. But part of the show's appeal, aside from its discussions about craft history, lies in the possibility that there could be something valuable and authentic. People watch to see if any sort of treasure might come to light, and the narrative of the show builds on that anticipation.

And over the years, audiences haven't been disappointed. Unearthing objects from a Shakespearean notebook to a Fabergé drinking vessel, from silver coins minted during the reign of Charles II to eighteenth-century Chinese carved rhinoceros horn cups, the *Antiques Roadshow* has given new life to some truly impressive antiques. (There was even a genuine Spanish Forger painting, discovered in 2016.) *Antiques Roadshow* runs counter to the stories of most fakes, frauds and forgeries – and that's part of its success. Instead of objects being debunked, these are stories that see items valued and authenticated, rare though these spectacular finds are.

What all fakes have in common is that they're usually too good to be true. Artefacts like the Spanish Forger's paintings, William Henry Ireland's Shakespeares or Johann Beringer's faux fossils seem to justify experts' scepticism – spectacular, earth-shattering, paradigm-shifting, unprecedented new discoveries in the art, artefact and antiquities worlds are assumed to be fake because, well, they generally are. But sometimes that default assumption is wrong. Sometimes new discoveries like those unearthed through *Antiques Roadshow* are just that – new discoveries. Such finds defy the odds, proving to be the Real Thing after all.

The ancient Maya Grolier Codex is an artefact with an odds-defying story. (Although the Codex has nothing to do with *Antiques Roadshow*.) For more than 40 years, many experts derided this ancient Maya artefact as a fake – the sort of thing that would fool an enthusiastic but undiscerning art collector – only to find after extensive scientific testing and decades of research that the codex is most likely the real thing. Although not universally accepted within the scholarly community, the Grolier Codex shows that the authenticity of objects is constantly under negotiation. It also shows that authentication and acceptance of something, after it has been declared fake, is an uphill battle.

★ ★ ★

To understand the Grolier Codex, it's necessary to unpack its history – the history of the codex's discovery, of course, but also the history of the ancient Maya scribes who wrote it, the Spanish who tried to destroy it and the decades of archaeological debate about its authenticity. Contemporary archaeologists put the creation and writing of the codex in the thirteenth century AD, and Maya history reaches back thousands of years before that.

Ancient Maya civilisation occupied much of southern and eastern Mesoamerica – from modern-day Chiapas in southeastern Mexico, through Belize and Guatemala, and into western Honduras and El Salvador – from several thousand years ago until the Spanish conquest of Mexico in the sixteenth century. In the millennia before the Spanish conquistadors ever set foot in Central America, many prominent Maya cities like Tikal, Palenque, Copán, El Mirador, Chichén Itzá and Calakmul enjoyed periods of political expansion, economic prosperity and ultimate decline as new cities emerged and others rose to power.

This history included a massive reorganisation of the Maya political landscape sometime around 900 AD, when centres of power shifted to the northern reaches of the Yucatán in Mexico and eventually to the volcanic highlands of Guatemala. With Spanish contact, the then-flourishing Maya city centres were 'crushed in a protracted, traumatic subjugation that consumed thousands of lives', prominent Mayanist Robert Sharer states in *The Ancient Maya*. The Spanish arrival 'was a scourge marked by brutality, catastrophic epidemic diseases introduced by Europeans, and the determined intervention of the Catholic Church'. Historians call this the Conquest and the Conversion.

When the Spanish arrived in the New World, they weren't prepared to encounter peoples with a history as long and complex as their own. Only their own cultural traditions with deep roots in the classical worlds of Greece and Rome, their Eurocentric logic went, were the legitimate route for a 'civilised' society to historically unfold. Mesoamerican cultures (particularly the Mexica, widely known as the Aztec) challenged their assumptions. Throughout their history, the Maya had demonstrated, in Sharer's words, 'astonishing achievements' in mathematics, calendrics, astronomy and, of course, writing, with complex technologies and complicated political organisations, and in the arts, with traditions in sculpture, painting and architecture.

In other words, 'to sixteenth-century Europeans, secure in the knowledge that they alone represented civilized life on earth, the discovery of the Mexica, the Inka, and the Maya came as a rude surprise,' Sharer explains in *The Ancient Maya*. 'The peoples of the Americas, though capable of brutal practices, were not as efficient in the practice of warfare as the Europeans, and although offering brave and determined resistance, were ultimately crushed by the conquistadors.' In the subsequent centuries, the idea that there was a 'lost civilisation of the Maya' resonated with Western audiences, which did not realise that there are millions of ethnic Maya peoples speaking dozens of Maya languages who, today, live in what we call Mesoamerica.

Consequently, the ancient Americas became something for mid- to late-nineteenth-century Europeans to encounter, explore and try to explain, and these Western outsiders categorically denied the living descendants of the ancient Maya any connection to this rich, intellectually and artistically sophisticated ancestry. But this historical aura was full of myths, legends and guesswork trying to make sense of Maya pyramids, architecture, ball courts (where sporting endeavours took place) and artefacts.

Nineteenth-century writers and travellers capitalised on this mystique. For example, between 1839 and 1842, American lawyer and travel writer John Lloyd Stephens and English artist Frederick Catherwood explored Mesoamerica and introduced their readers to 'lost cities' of the Maya, firmly establishing the Maya as exotic, extinct and curious to their American and European readers. Stephens and Catherwood took note of the Maya hieroglyphic writing throughout their travelogue, and 'believe[d] that [Maya] history is graven on its monuments', since they saw Maya writing on so many different buildings. But Maya writing was unreadable to Western audiences, as there was no

Rosetta Stone equivalent to translate the glyphs, leaving Stephens and Catherwood to ponder, 'Who shall read them?' As many Mayanist scholars point out and have for decades, deciphering the ancient Maya written language has been one of the most perplexing and enduring mysteries of the Maya.

Ancient Maya languages are written with a mixture of hieroglyphs that express whole words and others that connote distinct syllables. (To eighteenth- and nineteenth-century scholars, the script looked similar enough to ancient Egyptian to warrant referring to the writing as hieroglyphics.) Contemporary Maya languages, however, are most often written using the Latin alphabet, not glyphs, although there is a widespread push among indigenous groups to readopt the ancient Maya hieroglyphic system for ceremonial inscriptions written in various modern Maya languages. Very rough estimates suggest that something like 5,000 individual ancient Maya texts can be found in museums and private collections – most of which were written on ceramic vessels and stone monuments during the Classic Period (AD 200–900), although Maya glyphs also show up on cave walls, animal bones, shells and obsidian. (The earliest example of Maya hieroglyphic writing is a column of 10 hieroglyphs painted in a thick black line on white plaster, found on a 2,300-year-old limestone block in a temple in Guatemala.) These texts were written on incredibly durable materials, able to withstand centuries of weathering, thus offering modern Mayanists a written record to read and interpret.

Historically, the knowledge of how to read Maya hieroglyphic writing died out in the sixteenth century. Late nineteenth- and early twentieth-century scholars took up the question of translating Maya texts, and for the next hundred years linguists and archaeologists have worked to wring every possible bit of understanding from ancient

Maya glyphs. In their *Introduction to Maya Hieroglyphs*, Mayanists Harri Kettunen and Christophe Helmke describe the process of deciphering Maya texts as 'advancing … steadily in stages' over decades, punctuated by the occasional and spectacular breakthrough.

Maya glyphs are found on a plethora of durable artefacts such as ceramics and pottery, and architectural structures like temples and stelae, but codices were the primary form of written records. Made of sturdy paper from the inner bark of fig trees – called *huun* in Maya and *amate* in Spanish – the codices were folding books with continuous accordion-like pages, written by elite artist-scribes in hieroglyphic script in what experts consider to be the now-extinct Maya language, *Ch'olti'*. Maya script was used continuously from the third century BC until the Spanish Conquest, although some contemporary archaeologists contend that Maya hieroglyphic writing survived until the seventeenth century in areas that were unaffected by Spanish control.

Although parts of the various codices had been published throughout the 1800s, it wasn't until the latter half of the nineteenth century that linguists, epigraphers, archaeologists and anthropologists doubled down in their efforts to decipher the Maya script. While some experts were successful in decrypting the astronomical tables and the number system, full translations of the codices didn't happen until the late 1970s to the early '80s. As of 2017, scholars consider themselves able to translate roughly 90 per cent of Maya texts.

Maya book technology has always been codex-based. Broadly speaking, codices were handwritten books with pages that are bound together; Maya codices had folded, accordian-like page are a unique type of book and occupy an important place in the history of book technology. Globally, codices replaced scrolls – an earlier form of written records – almost completely by the sixth century AD, as a codex offered scribes a sturdy and compact set of writing surfaces and easily accessible information. Since both sides of a page could be used, and many pages could be bound together, scribes had more space to write than they did on just a scroll; additionally, to find something in a codex didn't involve having to, well, scroll through an entire book to find what one was looking for – bound pages could simply be flipped through. A codex, moreover, improved the durability and storability of a record. 'The painted leaves of codices were likewise read from left to right to the end of the front of the bark-paper strip,' archaeologists Nancy Kelker and Karen Bruhns say of Maya codices, 'then, if both sides were painted, the manuscript was turned and read from left to right so that the last page of the *verso* was the backside of the first page of the *recto*.'

Based on the existing Maya codices, scholars suggest that they are not records of historical events but are 'more esoteric and astronomical', as Mayanists Kettunen and Helmke write, full of 'information presented in the form of almanacs and prophecies'. According to codex translations, the Maya were particularly interested in Venus, as astronomer-priests had identified the Morning Star and Evening Star – that is, Venus – as the same planet. This Maya astronomical observation, contemporary archaeologist and Maya scholar Michael Coe was quick to point out, had not been appreciated by Homer's Greeks.

When Spanish conquistadors marched through Mesoamerica in the sixteenth century, the accompanying Catholic priests judged the Maya codices to be heretical, devil-inspired books and they burned them – nominally to demonstrate the 'superiority' of their Christianity over the polytheistic, nature-infused indigenous Maya religion. In July 1562, the infamous Spanish Bishop Diego de Landa congratulated himself on having incinerated the contents of an entire Maya library in the Yucatán. 'We found a large number of books in these characters [Maya script] and, as they contained nothing in which were not to be seen as superstition and lies of the devil, we burned them all, which they [the Maya peoples] regretted to an amazing degree, and which caused them much affliction,' he wrote in his *Relácion de las cosas de Yucatán*, his catalogue of the eradication of Maya history and literature.

Although some Catholic priests, like Bartolomé de las Casas, decried the destruction of the codices and had for decades, the extermination of Maya libraries was thorough – books were destroyed en masse. 'These books were seen by our clergy, and even I saw part of those which were burned by the monks,' de las Casas lamented in his *Apologética Historia de las Indias*, 'apparently because they thought [they] might harm the Indians in matters concerning religion since at that time they were at the beginning of their conversion.' The last of the then-known Maya codices was burned in 1697, when the Guatemalan city of Nojpetén fell to the Spanish. In a bit of historical irony, *Relácion de las cosas de Yucatán* also contained notes by de Landa about Maya script that would prove to be useful to scholars working to decipher and translate Maya hieroglyphs centuries later.

For decades, twentieth- and twenty-first-century Mayanists believed that only three Maya codices survived into the twenty-first century. The three of them – known as the Dresden, Paris and Madrid Codices – are named after the European libraries where they were rediscovered in the nineteenth century, when scholars began using the texts to study the natural history and ethnography of Mesoamerica. In 1811, for example, the famous naturalist-explorer Alexander von Humboldt published several pages of the Dresden Codex in his atlas detailing the indigenous populations of the Americas. These codices survived the destruction of the other Maya books only because Spanish conquistadors took them back to Europe as curiosities and souvenirs of their time in the Americas. Furthermore, these three codices all date to the later centuries of Maya book-making practices. Not a single codex from the Classic Maya Period survives today.

Ergo – and this is key to the story of the Grolier Codex – Maya codices are extremely rare, due to history, geography and chance. This scarcity impacts academic scholarship as well as the artefacts' financial valuation. Archaeologists can't just go out and discover more codices simply because they want to. Any *huun*-made book that survived the Spanish annihilation centuries before faces the threat of decomposition from Mesoamerica's wet, acidic soils, which easily eat through the delicate fig-bark paper. (Contemporary excavations of Maya burials have yielded some lumpy organic blocks with paint flakes that archaeologists suggest are the decomposing, unreadable remains of codices.) If codices were buried in dry caves, they could, archaeologists argue, survive for centuries. But no such codex has been recovered under any such circumstance, and the historical odds are that it never will be. Until the discovery of the Grolier Codex in the mid-twentieth century, the Dresden, Madrid and Paris Codices were it.

The rarity of Maya codices, however, piqued a very particular niche market in the early twentieth century, as their scarcity ensured a built-in demand in the world of high-end art collecting. This demand opened the collecting markets to artefacts that were looted from Maya archaeology sites and smuggled into auctions, and with this demand came a plethora of fake codices, crafted by forgers ready to foist their art into this burgeoning market. (The wealthy early twentieth-century American newspaper businessman William Randolph Hearst, for example, was said to have purchased not one, but two, fake Maya codices in the early twentieth century.) Although forged codices rarely fooled expert scholars who had seen the three authenticated codices for themselves, the fake Maya texts provided an extremely lucrative market for the less-than-discerning collector.

★ ★ ★

Faking Mesoamerican art – especially Maya art – is nothing new. Archaeological and historical estimates highlight this point, suggesting that Aztec forgeries were most likely sold to Hernán Cortés's unsuspecting, gullible soldiers as they conquistador-ed their way across Mesoamerica. These 'souvenirs' were essentially engineered artefacts for the Spanish, as enterprising Aztec artisans pivoted the Toltec statues and Teotihuacán masks that they had been forging in the past to include their new Hispano-Aztec market. (Some Mesoamericanist scholars think that it's possible that a pre-Conquest 'cottage industry' of forgers popped up around the ruins of Teotihuacán, to sell Teotihuacán fakes to the hunters of antiquity.)

The Spanish brought New World antiquities back with them, and many of the artefacts – some authentic, many not – were gifted across Europe from one noble to another, filling cabinets of curiosities with curios from new, exotic locations. The market for ancient Mesoamerican antiquities really took off, however, after the colonies gained independence from Spain in the nineteenth century. As the former colonies opened up for a plethora of businesses in the decades following their independence, demand for art, antiquities and artefacts from the 'lost civilisations' (which the Spanish had decimated during the Conquest) increased to the point where forgeries became incredibly productive. Travelogues like those of John Lloyd Stephens and Frederick Catherwood were powerful tools in the creation of interest in the mystique of the ancient Americas.

Moreover, earlier explorers and travellers like Alexander von Humbolt introduced European readers and the intelligentsia to the peoples and arts of the Americas. The developing European art market paired neatly with a rise in newly formed countries' interests in their indigenous pasts – in Mexico, for example, the War of Independence sparked political rhetoric that promoted politicians as restorers of the Aztec past. Other former Spanish colonies in Latin America began to look for ways to reclaim or promote their non-Spanish pasts, one of these ways being the construction of national museums that needed to be filled with artefacts.

From its earliest history, the faking of Mesoamerican art and antiquities affected both high- and low-end markets. Skilled craftspeople and artists ensured that whatever sort of piece a collector wanted – or expected – could be procured. And once fakes were 'authenticated' with the weight and authority of belonging to a collection, they were translated into 'real' artefacts, regardless of whether they were, technically, from the context they were purported to be associated with. 'By mid-[nineteenth] century there was an incredibly prolific and flourishing antiquity-fabricating industry,' Nancy Kelker and Karen Bruhns explain, 'at least in those countries that had significant numbers of foreign visitors/travelers/businessmen and had had complex societies producing "goodies" in the Precolumbian past.'

The situation with fakes, frauds and forgeries in museum collections reached a crisis point in the late nineteenth century. Alarming many museum professionals, like William Henry Holmes, the famous explorer-archaeologist and American museum curator. When Holmes took a curatorial position at the Smithsonian Institution's Bureau of American Ethnology in 1889, he was faced with the task of having to 'do something with' the large Latin American collection of arts and artefacts that the Smithsonian had acquired.

Holmes himself was an expert in the prehistoric Ancestral Puebloan culture (historically referred to as 'Anasazi culture'), and had worked with the US Geological Survey and, in 1883, had travelled to Mexico City to examine the archaeological collections at the National Museum. In his publications, Holmes noted that he found many forgeries in the National Museum, a situation he found perplexing. He was rather shocked, however, when he found copies of those same forgeries in the Smithsonian's collections! 'It is not surprising that the archaeologists in the United States or Europe should make mistakes in interpreting this work, as they have to take the word of unscientific collectors who rely upon the statements of native dealers,' Holmes clinically observed in his notes, 'but it is strange that Mexican scholars should so long have passed the work by without comment.'

It's worth noting that for many nineteenth- and twentieth-century collectors, the 'crudeness' of an early Americas artefact was considered virtual proof of its authenticity, regardless of its provenance or history. Cocooned in a rhetoric of cultural superiority and unencumbered by detail, many collectors simply assume that old-looking, stylistically 'crude' artefacts must be the real thing. Moreover, tastes in the then-contemporary art scene also drove demand, with artists like Diego Rivera collecting West Mexican figurines (and many fakes) because of their apparent 'modern art' flavour. It's also worth noting that 'Fakes are made for us. Fakers can appeal to the tastes of the modern era,' as art crime historian Erin Thompson reminded me when I asked her about the enduring appeal of fakes. 'The ancients didn't make things with our preferences in mind. Fakes are often bigger, more spectacular, more interesting, and more sexy than genuine antiquities.'

While all sorts of Mesoamerican artefacts were faked, and faked extensively, faux Maya codices have long been a class unto themselves. The first fake codices appeared on the market in the nineteenth century, overlapping with the burgeoning industry of other Mesoamerican fakes. In 1909, Mexican anthropologist and archaeologist Leopoldo Batres published *Antiguidades Mejicanas Falsificadas*, an extensive catalogue of Maya fakes, complete with photographs and descriptions of how the forgery industries operated. (He included pictures of brass stamps used by some forgers to easily crank out cheap, fast Maya-codex-looking artefacts.) Batres noted that forgers could acquire authentic-looking paper through shady archaeological site-looting connections.

For decades Mayanist and Mesoamerican scholars collected catalogues of codices that were understood to be phony – in 1935, Danish archaeologist Frans Blom published *A Checklist of Falsified Maya Codices*, listing the top 10 fake Maya codices. One of the codices that made it on to his list offered readers images of Maya warriors driving Roman-like chariots being pulled by plumed serpents. These forged codices were part of far-ranging museums and collections – one was in the Staadliches Museum für Völkerkunde in Munich and four were housed in Guatemala City. In 1958, prominent Mayanist César Lizardi Ramos denounced the then-prominent Codex Porrúa as a fake, bringing the number of fakes circulating within the museum and scholarly world to 33.

In their book *Faking Ancient Mesoamerica*, Nancy Kelker and Karen Bruhns argue that since Lizardi Ramos's work with the Codex Porrúa, the number of fake Maya codices has more than tripled, as of 2010. (And, of course, this estimate does not reflect the number of manuscript fragments and touristy souvenir-type codex forgeries.) 'The earliest published forgeries were unsophisticated works by modern standards but still sufficiently adept to pass the scrutiny of collectors,' Kelker and Bruhns point out. 'The more modern faux codices range from the unbelievably bad to those good enough to fool Ivy League Maya specialists.'

With the comfortable distance of decades, many of these debunked Maya codices appear ridiculous, even to the amateur eye. (The Roman-like chariots, for example.) Again, much like the work of the Spanish Forger and his nineteenth-century medievalism, once you know what to look for, the signs become easy to spot. Or, at least, they ought to become obvious enough to raise questions from buyers as well as connoisseurs. Many forgers worked off images from the authenticated Maya codices – the Dresden, Paris and Madrid – selecting motifs at will, but with no awareness or understanding of how and why some things ought to go together. They were essentially trying to imitate the Maya language without understanding syntax or context – and what they produced was simply a pictorial jumble.

One of the pages of the Codex Porrúa, for example, has 'whimsical pseudo-narrative scenes' that show a 'wedding-cake-like three-tiered pyramid' and a hiker who 'is in immediate need of a good podiatrist', as experts note in disgust. After wading through pages of mixed Maya and Mixtec motifs, also problematic for genuine Maya codices, the reader comes face to face with a dragon. 'The dragon,' Nancy Kelker and Karen Bruhns fairly smirk, 'came off of a Chinese import!'

★ ★ ★

Taken together, it's easy to see how – and why – authentication plays such an important role in Maya art, antiquities and archaeology. Which leaves us with the question: how do we know what's genuine and what's not?

First and foremost, if an artefact comes from a well-documented archaeological dig, the act of discovering it and recording that discovery – its provenience – helps to establish its legitimacy. When codices simply appear on the art market as 'recently discovered' but without proper provenience or provenance, it's impossible to tell if they're looted, smuggled, faked or in any other way unsavoury. (Across a plethora of disciplines, from art to archaeology to palaeontology, more and more experts are refusing to authenticate or even study objects for which the provenance cannot be verified. Private collectors' markets, however, are a different story.) Photographs, journals, site reports, monographs and academic journal articles – all of these things help provide a body of proof as to the genuineness and legitimacy of a discovery.

Aside from the context of an archaeological discovery – which may or may not be helpful in sorting through older collections – connoisseurship offers a qualitative expertise to vet pieces. Otherwise known as 'a good eye', connoisseurship is a go-with-the-gut sort of sense about judging whether paintings and other art are real or not, based on the experience of having seen and worked with art for decades. Nancy Kelker and Karen Bruhns remain a bit sceptical as to the efficacy of relying purely on connoisseurship as the best means to sift real art from fakes. 'Because of their appreciation of great art and their knowledge of all the juicy details about the lives of artists and collectors, connoisseurs make entertaining dinner guests,' they dryly note, 'but as a group, their track record detecting Precolumbian fakes is somewhat spotty.'

The mid-twentieth century, however, brought a plethora of new authentication methods to the worlds of art, antiquities and archaeology, as different scientific disciplines formalised specific types of chemical and physical analyses. These analyses offered means and methods for testing various characteristics of artefacts – information that could, in turn, be used as evidence for an object's authentication or not.

The first analysis available in 1949 was a systematic way to test the age of once-living materials – radiocarbon dating. This type of dating test only works on materials that were once organic; it destroys the sample being tested, and it takes a bit of skill to interpret the date range that the test gives. In other words, radiocarbon dating only works on things that were once alive, destroys part of the thing and doesn't just spit out a note saying, 'This codex was made in the year 1325.' But it was – and is – an important step in providing scientific analyses as part of the authentication process.

Since the mid-twentieth century, a plethora of scientific tests and methods has been developed. Thermoluminescence dating, for example, can be used on ceramics that have become divorced from their original archaeological context, offering a date for when a ceramic piece was fired. Computer-tomography (CT) scans allow researchers to 'see inside' an object. Scanning electron microscopes offer incredibly powerful magnification, leaving little room for forgers to hide any flaws in their craft. And these are but a few of the tools that science has to offer.

It's important to keep in mind that these tests cannot detect a 'real' object, and that there are many objects that defy any sort of scientific analysis. Sculptures, for example, are notoriously difficult to test. Tests can only tell us the age of an object, what it's made of and how it looks under magnification. The results of such tests have to be put together and read like a detective story. Consequently, analyses are generally used to debunk artefacts, by revealing results that show something was made centuries after it was 'supposed' to have been made, or something to that effect. Forgers have become more sophisticated over the years as well, adapting to the advent of these new technologies by trying to use looted ancient materials as a base for their modern creations.

'Tests can only detect fakes when we run the tests. And there are so many reasons why we don't run the tests.

They're expensive and it can be hard to find experts to run them. Often, we just don't want to run the tests. People or institutions who bought or sold an antiquity don't want to prove that it's worthless,' Erin Thompson told me in an interview. 'People who study antiquities don't want to prove that they wasted their scholarly career. The general public doesn't want to think that experts don't really know anything.'

★ ★ ★

Which brings us back to the story of the Grolier Codex. Although Mesoamerican archaeologist and Maya expert Michael Coe has championed the Grolier Codex as a genuinely authentic artefact from the beginning of the codex's story, it took more than 40 years of systematic study for other scholars to accept the accumulating evidence that it wasn't just some sort of fake, like so many other codices have been. And even today, acceptance of the Grolier Codex isn't universal.

There are few 'found' codices in recent Mayanist history that have been as contested as the Grolier Codex. It acquired its name from its first public showing in the Grolier Club in New York City, in 1971, organised by Michael Coe, then professor of Maya archaeology at Yale University. (The Grolier Club is a private society of bibliophiles with dedicated interests in the history of bookbinding, book printing and books in general.) The exhibition – 'Ancient Maya Calligraphy' – focused on intrinsically aesthetic objects that were either painted or inscribed with the Maya script, and Grolier Club members saw funerary vases, ceramic flasks, boxes and even a human bone that was incised with Maya writing. The codex, however, was the exhibition's centrepiece. Here, the exhibition claimed, was an artefact that defied historical odds.

The Grolier Codex features a Venusian calendar with Mayan glyphs and figures that are painted in red, black and blue over its white lime-coated stuccoed surface. The pages of the book are approximately 18cm (7in) tall and measure 125cm (50in) in length when they are folded out accordion-like – that's roughly the size of a contemporary paperback and, when stretched out, the length of a workbench desk. The codex comprises 11 surviving pages, but archaeological estimates suggest that it had as many as 20 pages. (The codex has five additional associated pieces of *huun* paper that have not been stuccoed or painted.) The bottom part of the codex is damaged as moisture has eroded and stained the remaining pages.

Although the codex has several rich colours – haematite red, deep black, and brown and red washes, as well as the famous 'Maya blue' hue – the inks have been used sparingly in the codex. One small part of the five associated pages was submitted to the Teledyne Isotopes laboratory for radiocarbon dating and returned a date of AD 1230 ± 130. If we assume that the five unpainted pages are of the same age as the painted ones, this puts the Grolier Codex as having been painted sometime in the thirteenth century, which Michael Coe argues is consistent with its style and content. The date would also, provocatively, make the Grolier Codex the oldest extant Maya book in existence. Not only was it unprovenienced, with odd calendrical information and strange blended iconography, but also Coe was claiming that it was *older* than the Dresden, Paris and Madrid Codices? Inconceivable!

Coe suggested that only one side of the Grolier Codex had been painted because the book was removed from circulation among the thirteenth-century Maya priests by either burial or ceremonially depositing the codex. When Coe and his colleagues described the codex's motifs and glyphs in the 1970s, they noted that there were some things, like the way that days are connoted in the codex, clearly comparable to the three other codices; however, Coe also suggested that there were some things about the style of the Grolier Codex that looked distinctly un-Maya and were reminiscent of other Mesoamerican styles from further west, like those of the Toltec and Mixtec. Scholarly reactions to Coe's claims were mixed, to say the least.

The Grolier Club published a catalogue with photos and descriptions of the exhibition's artefacts two years later, in 1973. In the entry for the codex, Coe described it as being from a private collection with unknown provenance. ('Said to have been found together with a mosaic mask in a late Maya-Mexican style … it must owe its preservation to the dry conditions of a cave somewhere in the Maya area,' Coe clinically notes. 'Its coming to light is thus an exceptionally rare event.') And this is where the story of the Grolier Codex gets complicated and its authenticity is challenged.

To begin with, the story of the codex's discovery didn't really make sense. Why would an artefact that rare have an unknown provenance, as Coe claimed? Why would a discovery this important – a fourth Maya codex – not provide audiences with the details of how it came to be exhibited? Journalist Karl E. Meyer pressed Coe for details about the codex's owner in Meyer's 1973 book about illicit art trafficking, *The Plundered Past*. Coe simply deferred the question and would only comment that, 'the codex was a "real hot potato" lent to the Grolier Club by an anonymous owner'. Just who found the codex, where, and under what circumstances was never very clear.

The story of the Grolier Codex's discovery as collected, pieced together and told by Karl Meyer, goes something like this: in 1966, a prominent antiquities collector in Mexico City named Dr Josué Sáenz received a phone call from a person who said, 'Hello. I will have good news for you in two weeks,' and promptly hung up. A few weeks later, the phone rang again and Sáenz recognised the voice of the man, who in this call identified himself only as 'Gonzáles' and promised to provide Sáenz with a cache of incredibly lucrative ancient treasures – Mesoamerican artefacts that would sell quickly and well within the antiquities market. Gonzáles called a third time and urged Sáenz to come and see the amazing treasures. Sáenz promptly booked a flight to Villahermosa in Tabasco, southern Mexico, no questions asked.

When Sáenz arrived in the foothills of the Sierra de Chiapas, he was met by Señor Gonzáles and other villagers, who showed Sáenz a spectacularly decorated Maya mask and the book with Maya hieroglyphics that would become known as the Grolier Codex, according to Meyer. Sáenz claims that the villagers told him that they had collected the two treasures from a dry cave at a nearby – but undisclosed – location.

A bit of background, here, about Dr Josué Sáenz. He was an imposing figure in the world of antiquities and art collecting. He attended Swarthmore College and the London School of Economics, and wrote a thesis about monetary theory while at Cambridge University, studying under prominent economic theorist John Maynard Keynes. Coming from an influential family in Mexico City, Sáenz had money, social position and connections. He was a banker, sportsman, public servant and even a university lecturer. (He was even on the National Olympic Committee during the Mexican Olympics in 1968.)

In the 1950s, however, Sáenz turned his attention and efforts to the world of collecting, quickly becoming a serious player in the world of Mesoamerican artefacts and antiquities. In *The Plundered Past*, one of Sáenz's colleagues described him entering the collecting world with confidence and money. 'Josué went about it the way he does everything, with intelligence and energy. He let it be known that he would pay the highest price on the market for anything he wanted, and as a result he got first refusal on everything of importance for sale in Mexico.'

This brings us back to Sáenz looking at the Maya mask and codex, standing on a rough airstrip in the Sierra Madre de Chiapas, surrounded by villagers. Because of Sáenz's reputation, Señor Gonzáles and his *colegas* would have known exactly who they were calling – the guy who could move their artefacts into an international art market and pay them for their troubles.

Sáenz is said to have offered the villagers $2,000 cash, then and there, for the two artefacts, but he coughed up more money after the villagers produced a catalogue from the Parke-Bernet Gallery in New York, and demanded that he pay them what would have been in keeping with such artefacts going up for auction on the art market. Locals – 'looters' in some archaeological literature, 'intermediaries' in other records – claimed to have removed the codex from a dry cave in Chiapas, and it was purported to have been found in a wooden box with three other pieces of *huun* paper and a turquoise mosaic mask. Sáenz asked to take the mask and codex to be authenticated before he bought them, but claims that the villagers refused to allow the artefacts out of their sight without payment.

Sáenz wrote a post-dated cheque, collected the mask and codex from the villagers, and flew back to Mexico City to have the artefacts authenticated by other experts. He hoped that post-dating the cheque would allow him to stop the payment if he found the artefacts to be fake. The villagers, however, quickly cashed the cheque and Sáenz was out of his money.

Authenticating the mask was an uphill battle for years, with some experts saying that it was real and others claiming that it was fake. (Sáenz himself had actually come to believe that the mask was a fake.) Eventually, the intricate mosaic mask made its way to Dr Gordon Ekholm, curator of Mexican archaeology at the American Museum of Natural History, who authenticated it by looking carefully at its construction, noting that it looked as though a piece of the mosaic had fallen off in such a way that it would have been a detail impossible to fake. The art world accepted it as the real thing and the mask was sold to American art collector Mildred Barnes Bliss in 1966; it remains in the family's private collection held by the Dumbarton Oaks Research Library and Collection, Washington, DC.

The codex, on the other hand, was much more complicated, and consequently its authenticity has been under debate for decades. Once Sáenz had purchased it, he tucked it away until he decided to show it to archaeologist Michael Coe in 1970–1971. It was at Coe's insistence that the Sáenz-owned artefact went on display at the Grolier Club with other examples of Maya writing.

In the business of separating real artefacts from fakes, first impressions are everything. A provenance is an integral part of any artefact's cultural history, as explanatory power and cultural cachet are largely predicated on the circumstances that surround its discovery. How an artefact is discovered – and how that discovery is documented – is just as important to the process of authentication as the physical properties of the thing itself. The story of an artefact's discovery is its introduction to the world, and those initial impressions are hard to shake off; if there is anything suspicious, the whispers of fakery or forgery will follow an artefact for the rest of its life, directly affecting its social, ethical and financial valuations.

If the original context is legal and legitimate, artefacts are more readily accepted – they quickly move to being material objects to be studied by academics, accessioned into museums by experts and bought by private collectors, depending on an artefact's legal ramifications. On the flip side, looted or forged artefacts are often backstopped with fabrications so vague that they are impossible for anyone to falsify. If the documentation of a discovery cannot be proved one way or the other, the default assumption is that an artefact is simply not the real thing. In the decades since Sáenz acquired the artefacts, his story has rankled many an expert.

Just about everyone who was anyone thought the codex was a fake. From the start, the unverifiable provenience (and the story that smacked of looting at best and forgery at worst) meant that authenticating the artefact would be incredibly difficult. Furthermore, for over a century, Maya experts had become comfortable with the idea that only three Maya codices had survived the Spanish Conquest, so the possibility that a fourth would so conveniently turn up, when all other discovered codices had proven to be fakes, strained credibility. Assuming that the Grolier Codex was a fake was an easy – and justifiable – assumption. It made it difficult for audiences to take it seriously as a legitimate new discovery. Even decades later, the story of its discovery is vague at best, sketchy at most, and quite frankly comes across as Hollywood-inspired fiction. What, everyone asked themselves, were the odds?

In the decades after its initial exhibit, experts debated, scientifically tested and questioned the Grolier Codex, finding a plethora of things to seize upon that indicated the artefact wasn't the genuine thing.

In his seminal study of the codex in 1976, archaeologist J. Eric Thompson built a powerful case to dismiss the artefact as a fake. In addition to the swarmy story of its discovery, archaeologists pointed out that that the Grolier Codex had several, significant differences from the Dresden, Paris and Madrid Codices – everything from the number of blank pages, to the style and literary scope of the text, and the fact that it was purportedly older than the other three. And just because one piece of paper associated with the codex was radiocarbon dated to the thirteenth century, his argument went, it didn't mean that the rest of the codex was that old.

Moreover, Thompson pointed to the longstanding tradition of forgers using bits of genuinely old paper and decorating them with Maya symbols to fetch a better price for an object. There was the question of the uncharacteristic motifs and glyphs that Coe had noted in his original catalogue in 1973; they weren't in keeping with how archaeologists thought about the Maya in the mid-1970s. Some scientists claimed that the sharp edges along some of the pages could be explained by a modern blade cutting the paper, and that the water damage was a convenient cover-up to offer the purported Grolier Codex forger enough wiggle room to get away with Maya script that wasn't textbook correct. Other experts focus on the 'freshness' of the paint (compared with the three authenticated codices), and suggest that it would only be that good if it was a much more recent fake.

The three authentic codices are all books that show divination or prophecy – the Grolier Codex does not. Other experts do not believe that the motifs and glyphs actually illustrate Venus, thus undercutting the Grolier's potential authenticity. The Dresden Codex, for example, displays several calendrical cycles of Venus as well as depicting five different gods of the morning star, each positioned in the centre of a separate page displaying the dates associated with the four phases comprising each Venus cycle. Where the Dresden Codex shows one Venus god representing the morning star in each of the five cycles of Venus, according to the Maya calendar, the Grolier Codex shows four Venus gods in each cycle – a noticeable departure in motif.

Over the decades, experts have been torn as to whether or not the variations in the calendar notations meant that the author of the Grolier Codex was familiar with the notation and ring numbers found in the Dresden Codex. 'I sincerely doubt that any modern faker would have thought of putting hybrid ring numbers into a Venus calendar,' Coe dryly suggests in his original 1973 report, when whispers of doubt about the artefact's authenticity swirled around the artefact and its sketchy provenance. 'Fakers, whose knowledge of the Maya calendar and iconography is fairly abysmal, are usually reduced to copying, but no trace of copying [from the Dresden Codex] can be detected here.'

And so some archaeologists, like Michael Coe, said that yes, the Grolier Codex was authentic; others, like Nancy Kelker and Karen Bruhns, said that it wasn't. However, it feels telling that in Kelker and Brunhs's *Faking Ancient Mesoamerica*, the Grolier Codex gets six pages of discussion about just how it is fake, while other fakes are debunked succinctly in the space of a page or so. (They let that afore-mentioned imported Chinese dragon motif, for example, basically speak for itself.) But their nuanced treatment of the Grolier Codex highlights its ambiguity, suggesting that the artefact is complicated enough to warrant much more discussion than has been generated for any other purported Maya codex. In short, for decades it was simply easier to find legitimate ways to discount the Grolier Codex than it was to accept it as authentic.

Michael Coe and a few other archaeologists, however, were willing to bet their expertise that what they had seen defied the historical probabilities and was the real deal. Building a convincing case against fakers – imagined or not – would take time, and teams of archaeologists and Mayanist scholars spent that time gathering evidence. In the two decades since its discovery and initial exhibit, archaeologists, art historians and epigraphers had not been idle. Today, 40-plus years of studying the Maya networks and influences have helped to explain the odd Toltec and Mixtec motifs that Coe noted in his original description. (It turns out that there was a lot of variation in how glyphs were written – differences that depended on where in the Maya empire the script was found, as well as, more significantly, *when* it was written.) Motifs that had perplexed experts now had solid explanations.

In 2007, scientists published a detailed analysis of the so-called Maya blue pigment typical in painted Maya script and found in the Grolier Codex. The idea was to see if the blue found in the Grolier Codex matched the materials of other, validated Maya blue sources. The chemical make-up of the blue pigment is particularly important, furthermore, because at the time of the suggested illicit fabrication of the Grolier Codex, no one knew how 'Maya blue' was made – it would have been virtually impossible for a forgery to have the correct chemical make-up of true Maya blue, because it had not been revealed itself.

Maya blue is a vibrant, sky-coloured azure colour found on ceramics, architecture and written records across the ancient landscape of Mesoamerica. First created around AD 300, Maya blue melts together indigo from the local *añil* plant and the clay mineral palygorskite to form an ink that has endured in the archaeological record for centuries. According to archaeologists, Maya blue was ritually made by heating together the palygorskite and indigo in incense burners. More than just something to write with, however, Maya blue was a critically important part of ancient Maya religion and ritual as it symbolised the rain god, Chaak, as well as being associated with other deities.

If the Maya blues in the other, authenticated sources of blue match those in the Grolier Codex, the logic went, this would help to shore up the argument about the codex's authenticity. Using non-destructive methods like PIXE (Proton-Induced X-ray Emission) and RBS (Rutherford Backscattering Spectrometry) analyses, scientists examined the elemental make-up of the Grolier Codex's Maya blue pigments to see if any modern stuff had made its way into the paint's make-up; results from the study indicated that there wasn't anything modern in the pigments (meaning that the paint would, in fact, be very old), and that the array of minerals and elements found in the Maya blue was consistent with other archaeological discoveries. 'We are however a bit closer to determination of its authenticity,' the study authors noted, 'but other factors must be considered, such as deterioration patterns, content and context. Materials analysis is just one of the facts.'

The scales tipped in favour of accepting the Grolier Codex's authenticity in 2016, when Stephen Houston of Brown University and Michael Coe (now professor emeritus of archaeology and anthropology at Yale), along with Mary Miller of Yale and Karl Taube of the University of California-Riverside, published a peer-reviewed synopsis of the decades of study of the Grolier Codex. Each brought their own expertise and academic specialisation to the project, and the goal was to put to rest the question of the codex's authenticity.

They noted that the sharp cuts on the codex's pages are not the result of modern tools (as earlier experts had argued), but of a natural pattern of breaks in the gypsum-based stucco plaster used to coat the pages; they also argued that the sketch and grid lines in the glyph writing are in keeping with how glyphs in Maya murals were painted. And, finally, additional radiocarbon dating of the paper gives a date range of AD 1212 ± 40, corroborating the original dates of AD 1257 ± 110. 'Our goal was to see if there was something that was modern that [may] have been put into the paint on the Codex … to confirm that it was a 20th century fraud,' Yale

art historian Mary Miller explained in a 2016 interview
with *PRI.org*. 'Having started in the corner of the doubters,
I have moved to the opposite side of the ring and I have no
doubt of its authenticity.'

'It became a kind of dogma that this was a fake,'
archaeologist and Mayanist scholar Stephen Houston
declared upon the publication of what has been seen to be
the Grolier Codex's authentication. 'We decided to return
and look at it very carefully, to check criticisms one at a
time. Now we are issuing a definitive facsimile of the
[Grolier Codex]. There can't be the slightest doubt that the
Grolier is genuine.'

Taken together, the conclusion is well reasoned and
consistent: the Grolier Codex is the oldest known codex
created in the Americas – a record of astronomy and
calendar keeping from the late Maya civilisation. For all
intents and purposes, it's the Real Thing.

★ ★ ★

This is an unusual direction for a story about authentication
to take. Instead of experts debunking artefacts as forgeries
or fakes – like the Spanish Forger's panels, William Henry
Ireland's Shakespeare signatures or Beringer's faux fossils –
a small group of Mesoamerican scholars had to work for
decades to provide a convincing argument to their
colleagues and to the world that the Grolier Codex was
genuine.

But debate still surrounds the Grolier Codex is far
from over. In the August 2017 issue of the scholarly
publication *Mexicon*, Mayanist Bruce Love argues that
the scientific analyses associated with the codex aren't
definitive enough to justify the interpretation made
around them. He argues that the Maya blue isn't a positive
result, but rather simply isn't inconsistent with other

negative results. The water stains on the pages, Love maintains, don't penetrate below the gypsum overcoat of paint the way they ought to in an artefact that had been exposed to moisture. And then there's the issue of the calendric cycles, iconography and glyphs that have never matched expectations. 'In my opinion,' Love argues, 'the authenticity of the Grolier will never be decided by iconography, which can be argued *ad infitum* ... we can only hope that the Grolier Codex will undergo another round of scientific testing.'

Much of the problem of the Grolier Codex's authenticity comes back to the question of its origin – because its provenance is so problematic, some experts think that it's impossible to judge it as authentic or not. Because it wasn't recovered in an archaeological context – it wasn't excavated or mapped – there have always been questions about just how much of Dr Josué Sáenz's story is legitimate and how much is simply an embellished and sanitised way to talk around the possibility that he bought looted artefacts and then helped traffic them to the United States.

'Sáenz was a controversial figure,' Mary Miller said in a 2017 interview with *Yale News*. 'People disliked his involvement with the 1968 Olympics. People resented the fact that he was a private collector who didn't donate his collection to the nation. They disliked the man and his collection, and they wanted to treat everything in it as fake.'

Donna Yates describes that process of creating a false provenance as a common way to make illegitimate artefacts saleable in the art and antiquities markets. 'Sellers make up plausible but totally fake stories that explain why the objects are in their possession, and which render the pieces sellable on the market,' Yates explains. 'No one actually believes these stories, but the better ones are just tight enough that they are hard to prove false. The lowest level is saying that a recently-looted piece comes from an 'old family collection' or an 'Anonymous Swiss Collector'. Even in the most unlikely of situations, these 'false provenances' can win out.'

Consequently, the context of an artefact's discovery sets the null expectation for its authenticity. With a good origin story of a properly papered and provenanced discovery, the expectation is that the artefact is genuine; but a story that smacks of ill-repute immediately sets expectations that the article in question is forged or a fake. Even now, after the Grolier Codex's authentication, there are experts who don't embrace the codex wholeheartedly, pointing to its probably looted and possibly trafficked provenance. It's a morally and ethically inconvenient genuine artefact, significant though it may be.

'An archaeological find of major significance was made at the site to which Dr Sáenz was summoned in Chiapas,' Mayanist Clemency Coggins lamented back in 1973, lambasting colleagues who were willing to study artefacts without clear provenances. 'Whatever was in the cache has been dispersed, and the item of most importance [the Grolier Codex] has ... whereabouts uncertain. Such is the high price we pay for the high price of art.'

After its display at the Grolier Club in New York City, the codex was repatriated to Mexico City and has been housed in the Museo Nacional de Antropología for decades, with a replica of it on display for museum visitors. The story of the Grolier Codex is a reminder that objects live on a continuum of authenticity, and that they can move up and down that continuum, depending on their history and context. 'Authenticity is cultural. What we consider to be "real" depends on our social and cultural circumstances, our world view, and the context that we are in,' Donna Yates described to me in an email. 'Stories create authenticity and, indeed, authenticity is a story.'

As for Michael Coe and his lifetime's work with the Grolier Codex: 'It's nice to be vindicated,' Michael Coe said in a recent press release about the codex's authenticity, adding, 'I never once changed my mind in these 43 years. I knew it was good.'

CHAPTER EIGHT

The Art of Making the Palaeolithic Come to Life

It's hard to know just which piece of art in the exhibition's gallery to focus on first.

On one side there's a painted panel of horses with their grey coats and black manes fading in and out, offering visitors the illusion of movement. Further along the wall there is a mosaic of red dots arranged in an abstract pattern. Multiple illustrations of now-extinct rhinos (from the genus *Megaceros*) line the gallery walls. An owl, its unpainted outline traced with the artist's fingers, stares out at visitors, and warm-coloured lights showcase the gallery's underlying yellows and oranges. At the end of the gallery visitors encounter a 12m (39ft) long fresco of lions. There are hundreds of Pleistocene-age animals in the exhibition, representing something like 15 different species. Photographs are not allowed.

Welcome to the Caverne du Pont d'Arc – a replica of the Palaeolithic Chauvet Cave, a UNESCO World Heritage Site. The gallery is cool, dark and feels just a little bit clammy – very much like a cave. It's a cross between an art museum, a living history exhibition and the world's most expensive copy of an archaeological site.

For most, the replica of Chauvet is as close as they will ever come to setting foot in the cave.

★ ★ ★

On 18 December 1994, French cave specialists – speleologists – Jean-Marie Chauvet, Eliette Brunel Deschamps and Christian

Hillaire discovered an extraordinary Palaeolithic cave in southeastern France. The discovery was made as part of a survey of the Ardèche region's caves. This particular one overlooked the River Ardèche near the Pont d'Arc, where the meandering river had cut out an arch from the surrounding limestone – itself an iconic bit of French geography. The first of the three cavers to squeeze through the cave's small tunnel entrance, Brunel Deschamps, shouted out, 'They have been here!'

'They' were *Homo sapiens* from the end of the last Ice Age, people who lived in Europe tens of thousands of years ago. The cave that Chauvet, Brunel Deschamps and Hillaire had discovered, like so many other Palaeolithic caves in France, was full of artefacts and fossils that contemporary archaeologists use to piece together Palaeolithic life, one artefact, one painting, one site, at a time. But something about this one was different.

This particular cave had footprints of the people who had been in the cave millennia before. There were hearths and the remains from fires last lit during the Pleistocene, and knapped pieces of flint, from people making stone tools, were scattered around the floor. There was even a projectile point made of mammoth ivory, some 30cm (12in) in length, found later near one of the cave's hearths. A step had been deliberately made in the cave's floor, in one of the cave's high-traffic areas. The lower parts of the first chambers were carpeted with bones, as the skeletal remains of animals had been transported there by water for tens of thousands of years. (Archaeologists would eventually catalogue more than 170 bear skulls, from the now-extinct *Ursus spelaeus*, in the cave.) Tracks from bears, dogs and even an ibex crisscrossed the different chambers. Stalactites dripped from the roof of the cave and calcite deposits continued to grow on all of the cave's surfaces as they had since the Pleistocene.

But most exciting of all – and what separated this discovery from the hundreds of other Palaeolithic caves like it – was the cave's phenomenal, breathtaking art. This art was what Brunel Deschamps saw that first December afternoon. These were not just a few pictographs or an isolated painted panel, like ones the speleologists had found at the nearby archaeological sites of Chabot, Le Figuier, Olen, La Tête-du-Lion, Les Deux-Ouvertures and Ebbou. This cave had hundreds and hundreds of pictures that spanned thousands of years.

The walls were painted with animal and human images over two distinct periods, separated by roughly 5,000 years – the first 30,000–32,000 years ago and the second 26,000–27,000 years ago. (The question of the cave's dates has been highly contested, with some archaeologists arguing that the dates were much too old for art this sophisticated. Subsequent dating of charcoal and even some of the faunal remains from the cave has confirmed the date ranges.) Lions, mammoths, rhinoceroses and cave bears make up the majority of painted animals, but others feature as well. Some panels feature red dots and the stencilled outlines of handprints. For more than 20,000 years, the cave was undisturbed – its art and its artefacts just as they were when the last Palaeolithic artist left the cave. 'Suddenly, we felt like intruders,' the speleological team reflected, reverently taking in the galleries of art that surrounded them. They were not the first to be there.

It's not easy to discover an archaeological site – it involves both skill and luck. And to a large extent that luck and skill depend on an area's landscape – whether a place's geography was conducive to ancient peoples using it long ago, and if that same place can be discovered as an archaeological site today. Landscapes are not static, and how a cave looks to us today isn't necessarily how it would have looked to people 30,000 years ago. Due to its karstic geology, the Ardèche region where Chauvet, Brunel Deschamps and Hillaire spent their time caving is honeycombed with limestone *grottes* where water has weathered away the cliffs' dark grey rock, leaving hundreds of caves behind. For years, the three had explored this area; they and countless other experts had walked by this particular cave without suspecting that it was even there.

Even when explorers think they've found a cave, accessing it can sometimes be the most difficult part. Many ancient entrances to caves have been filled in or blocked over time, and it requires speleologists with extensive caving experience to seek out alternate ways to get in. To find new entrances, the team of cavers would test for air passing through scree and rubble around the cliffs – such draughts were good indicators of a possible way into a cave hidden well beneath the ground. ('Jean-Marie likes to use the back of his hand, while Christian and Eliette prefer to expose their faces to them,' they note in their monograph about the discovery.)

Once inside a discovered cave, it is long and tedious work to clear ducts and to systematically follow passages. 'If we find archaeological material, we leave absolutely everything in place so that the scientists can study the site exactly as we found it, and sometimes just as prehistoric people left it several thousand years ago,' the speleologists explain in the introduction to their book, *Dawn of Art: The Chauvet Cave*. 'If no protection can be ensured, the objects are collected and deposited in the regional museum of prehistory at Orgnac.' But they are quick to point out that finding art – that is, Palaeolithic art – is certainly the exception rather than the expectation for any sort of archaeological survey, including theirs.

Around three o'clock in the afternoon on the cold, wintry Sunday of 18 December 1994, the cavers followed an old mule path that offered a spectacular view of

Pont d'Arc, until they discovered a narrow opening, about 80cm (31in) high and 30cm (12in) wide, and all of them wriggled through it. Once inside, they could just barely stand up, but the cave soon opened up to massive chambers in which they could easily walk around. As they walked further in, they knew they would need ladders and other equipment to safely explore the entire cave with all of its chambers. It was the cave's art that was truly exceptional. Nowhere else in France – indeed, nowhere else in the world – had Palaeolithic art like it been discovered.

After their initial exploration that evening, the cavers carefully re-blocked the narrow entrance passage with stones to ensure that no one could enter the cave and cause any sort of damage to the art. They ended up going back later that same night to show the cave to Brunel Deschamps's daughter; over the next week, they revisited the site with another group of three cavers. 'This first visit had lasted only an hour, but we were in a state of shock because of our discovery,' they recall in *Dawn of Art*, 'both moved and, in a way, crushed by the weight of such a responsibility.'

In their subsequent visits the explorers carefully moved from painted panel to painted panel, taking in what they saw and photographing the art – everything from horses, hand stencils and red dots, to rhinoceroses and reindeer. The sheer volume and enormity of the discovery began to sink in, as did the responsibility they felt for the preservation of the cave. Having surveyed hundreds and hundreds of caves in the region, and finding ones with art to be very rare, they were more than primed to understand how to care for the art in the cave and to introduce conservation measures as soon as they stepped into the dark limestone chambers and saw what they'd discovered.

The cave – that came to be known as Chauvet Cave – quickly became an icon of Palaeolithic art, one of the most significant sites discovered in over a hundred years of palaeo-research in France. On 22 June 2014, Chauvet was put on UNESCO's World Heritage List.

★ ★ ★

Before Chauvet could become iconic, its paintings had to be authenticated. While Palaeolithic fakes are rare, they are not unheard of. Consequently, after the cave was discovered, the first step towards its preservation and scientific study was proving that its paintings were genuine Palaeolithic art.

'I was travelling abroad on 18 January 1995 when the news of the discovery of Chauvet Cave hit the media, and my initial reaction on seeing the first pictures of the amazing rhinos and big cats was that this might be a fake,' prominent British archaeologist Paul Bahn recalled in *Return to Chauvet Cave*. 'They were simply too stunning and too unusual to fit the picture we have built up of Ice Age cave art over the past century. But as soon as I learned who had found the cave, I realized that it must be the real thing.'

The discovery of Chauvet Cave was potentially so significant that the Ministry of Culture was able to pry Jean Clottes, the renowned French Palaeolithic archaeologist, away from his family's Christmas celebrations the following week to ensure that the cave was immediately vetted properly and that official conservation measures were outlined. The authentication couldn't even wait until after the Christmas and New Year holidays because the question of preservation and conservation of the paintings was so pressing. On 29 December 1994 – just 11 days after the cave's discovery – the three cavers met with Jean Clottes, Jean-Pierre Daugas, regional curator of archaeology, and Bernard Gély, who was in charge of DRAC, the regional archaeology service for the Drôme, Ardèche and Isère areas.

These officials were all there to authenticate the cave's paintings.

Before actually shimmying down into the cave, Jean Clottes was mischievously sceptical, suggesting to the group that he fully expected to see some fakes; in other words, what Chauvet, Brunel Deschamps and Hillaire claimed to have discovered defied the then expectations of Palaeolithic cave art. Brunel Deschamps reassured Clottes that when they all left the cave that afternoon they would want to drink the champagne she had brought to celebrate. And celebrate they did, as it was quickly clear to all the experts that the motifs and paintings were the Real Thing. The initial authentication has been upheld through decades of study and a plethora of radiocarbon dates, putting the paintings' charcoal pigments in the two different late Pleistocene periods.

It turns out that Palaeolithic cave art is actually rather hard to fake, or at least, it's hard to fake well. Historically, phony ice age artefacts crept into museum collections and art markets as soon as real Palaeolithic artefacts began to be discovered in Europe in the 1860s. In south-west France specifically, Palaeolithic artefacts that were easy to dig up and carry away – like carvings – popped up in the art market for over 150 years, at a time when labourers were often paid by the find and systematic, scientific methodology was scant. (Some of the nineteenth-century excavations were, for example, routinely 'supervised' without even bothering to be on site. This not-uncommon practice has led to problems in sorting out proveniences of artefacts, casting doubt about the legitimacy and interpretation of several significant finds from digs.) Many of the portable pieces of art that were carved from materials like stone or genuine fossil ivory were, and still are, impossible to authenticate as genuinely ancient. With so little control over an artefact's provenience, forgers found themselves with a new, lucrative scheme, and copies of Palaeolithic artefacts began to infiltrate that particular niche of the art and antiquities markets.

Faking or forging paintings on cave walls, however, is much more difficult to pull off. Unlike other forgeries of portable Palaeolithic art, one can't simply conjure up a cave the way one can source other media; nonetheless, faking cave art, however, isn't unprecedented. As early as 1909, Palaeolithic archaeologists Henri Breuil and Jesús Carballo examined some cave paintings in Las Brujas Cave in Spain's Cantabria region – an area with well-documented, authentic Ice Age cave art – and determined that they were out-and-out hoaxes. The styles didn't match other Pleistocene art in the region, the motifs were wonky – all the signs in fake art that tip off experts were present in Las Brujas. (These fakes were destroyed in 1960.) Over the course of the twentieth century, other more convincing faux cave paintings took to intermingling fake motifs with the real thing – say, introducing a bison into a genuine Palaeolithic scene where there wasn't one before. But these are rare.

Perhaps the most famous example of faked Palaeolithic cave paintings occurred at Zubialde, a cave in the Basque region of Spain. When photographs of the cave's art hit the European press in March 1991, 'most specialists immediately smelt a rat', as Paul Bahn put it in *Journey Through the Ice Age*. To begin with, the animals featured in Zubialde's paintings – rhinos and mammoths – were highly unusual motifs for Spain. The bison were 'very ugly and clumsily executed', as were the stencilled handprints. But most problematic were the parts of the paintings that suddenly appeared in the time between the cave's initial photographs by the young 'discoverer', amateur speleologist Serafin Ruiz, and when specialists examined the paintings. The Zubialde forger, suspected to be Ruiz, did not use carbon-based pigments (which could be directly radiocarbon dated) in the cave's paintings. However, subsequent scientific analyses of the paints showed that they were modern – they contained highly perishable materials like insect legs, biotic material that would never have survived from the Pleistocene to today. The pigment also contained synthetic fibres from modern kitchen sponges.

It took long and very detailed studies by Basque Palaeolithic specialists Ignacio Barandiarán, Juan María Apellániz and Jesús Altuna to discredit the Zubialde hoax, and before the cave's paintings were debunked they received a lot of media attention. Because of fakes like Zubialde, when caves with spectacular or unexpected motifs, like Chauvet Cave, are discovered, there is more onus on them to be proved genuine.

A few months after the Zubialde episode, a different Palaeolithic site, Cosquer Cave, was discovered in 1985 along the southwestern coast of France. The entrance to Cosquer Cave is now around 37m (121ft) underwater due to how much sea levels have risen since the Pleistocene, and archaeologists must actually dive to get into the cave, where intricate cave paintings were found above the water levels. When the first photographs were published in 1991, they strained archaeological credibility – the site included epically spectacular Palaeolithic art, with unusual animals like great auks, as well as other motifs. It was hard to believe, as Paul Bahn recalled, that the cave's art was authentic 'with its unprecedented geographical location, its entrance beneath the Mediterranean – so reminiscent of the plot of Hammon Innes' novel *Levkas Man* – and its unusual drawings'. With high-quality photographs, however, doubts about the legitimacy of the Cosquer's paintings were quickly put to rest. It was bizarre and beyond unexpected, everyone agreed, but it was authentic.

Less than three years after the discovery of Cosquer Cave and the Zubialde incident, France's Ministry of Culture

announced the discovery of Chauvet and released a series
of photographs of some of the cave's art. Consequently,
archaeological audiences were primed to be sceptical, but
Chauvet was the real deal. 'For all sorts of reason, primarily
a conservative tendency among us all when faced with
the new, every great discovery has aroused its share of
controversy and doubt,' Jean Clottes offered in the first
extensive publication of Chauvet Cave's art. 'It happened
with Altamira at the end of the last century, Lascaux in the
1940s, Rouffignac in the mid-1950s, and Cosquer in 1991
and 1992. Only Chauvet Cave seems to have escaped these
suspicions.'

With sites like Zubialde and Cosquer in mind, Clottes
joked about expecting to find fakes before he saw Chauvet
painted panels – but the authenticity of the cave's art has
never been in doubt. The original cavers took too many
precautions and were simply too good at what they did to
have their discovery called into question.

★ ★ ★

Chauvet Cave extends for something like 500m (1,640ft).
A succession of four chambers contain painted and
engraved figures, and ceilings stand anywhere from 15m
to 30m (49–100ft) tall. In the two weeks between the
discovery of Chauvet Cave and its authentication, the
original cavers took precautionary measures to ensure that
the cave remained as undisturbed as possible, to preserve the
integrity of the site and to prepare it for archaeological
study. They put down hundreds of metres of black plastic
sheeting – about 50cm (20in) wide – wherever there might
be a walkway, and marked off delicate areas with fluorescent
ribbons to keep from treading on fossil teeth, skulls, bones
and Palaeolithic hearths. The cavers walked in single file,
often in their socks to limit their own footprints in the cave.

After the cave's art was authenticated, and before it was introduced to the world, officials wanted to make sure that the site was secured. The entrance to the cave was on private property, located in the commune of Vallon-Pont-d'Arc. Once the landowner was informed of the cave's discovery, he was agreeable to installing a gate to guard the entrance and help limit the curious from traipsing through the cave.

On 14 January, 27 days after the discovery of the cave, the three speleologists got the go-ahead from the French government and, with the help of Monsieur Brunel, Eliette's father, made a high steel door, roughly 1.25m (4ft) high, to block the limestone opening, having carried sand, water and cement to the entrance to put it in place. The entrance used by the three cavers was roughly a foot tall, slanted downhill and went on for something like 7m (25ft). Photographs of cavers using that entrance to Chauvet Cave show the soles of their shoes disappearing through the opening with very little space left around them. (They entered the extremely narrow cave chute head-first). That entrance has since been widened and is guarded by the door. The entrance used by the Palaeolithic artists is now blocked by scree, while the original Pleistocene entrance was blocked sometime before 12,000 years ago. Everyone took care to make sure that the entrance had the same ventilation as it had before its discovery to avoid changing the cave's equilibrium and causing damage to the cave's art.

For the next couple of days, French gendarmes watched the cave day and night until electronic alarms and video surveillance were installed along with the door. On 18 January 1995, the team held a press conference organised by the Ministry of Culture, at which minister Jacques Toubon made the official announcement of the cave's discovery. And then all science and heritage conservation broke loose.

Following the announcement, other conservation measures were undertaken over the next year to facilitate scientific study of the art and archaeology of Chauvet Cave without destroying it. Narrow metal walkways were installed throughout the cave, taking care to use the smallest area possible. The walkway between the cave's Gallery of the Megaceros (the panel with paintings of the extinct Irish elk) and the gallery at the very back of the cave was equipped with handrails for navigating down a tricky incline. Electric cables were installed. A multi-year interdisciplinary archaeological study was commissioned, headed by Jean Clottes.

While archaeologists would be allowed to undertake scientific studies, systematically mapping the cave, documenting its art, and collecting charcoal and pigment samples, it was abundantly clear that Chauvet Cave would be open to scientific study only under strict supervision and that the cave would never, ever be open to the public.

★ ★ ★

Concerns about limiting access in order to preserve Chauvet Cave's paintings were well founded, based on the destruction of Palaeolithic art in Lascaux Cave half a century before.

Lascaux Cave was discovered in 1940 by local teenager Marcel Ravidat (and, purportedly, his dog), soon after France fell to the Axis invasions. More than 400km (250mi) west of Chauvet Cave, Lascaux is a natural limestone cave located on the left bank of the Vézère River in the Dordogne region. Smaller than Chauvet, Lascaux is roughly 250m (820ft) long and has more than 6,000 images of human figures, animals and abstract symbols painted in reds, yellows and blacks. The images that wend their way along the cave walls date to 17,000 years ago.

The cave's most famous panel is the Hall of Bulls, depicting 36 different species, including one 5.2m (17ft) long bull that appears to be in motion. After supposedly visiting Lascaux, Pablo Picasso is said to have announced that 'We have learned nothing in twelve thousand years.' (Archaeologist Paul Bahn thoroughly investigated Picasso's rumoured quote, concluding that yes, Picasso's paintings, like the lithograph series of *The Bull*, resembled Lascaux's bovid-based motifs. However, there is little to no evidence that he actually visited Lascaux or ever physically stood in awe of the Pleistocene's artists. But, like so many things, the story has taken its own place in history.) Regardless of whether or not Picasso actually ever commented on Lascaux's art, just the rumour of it helps cement the site's modern cultural cachet.

In the 1940s, the entrance to Lascaux was enlarged considerably, making it easier for visitors to navigate their way into the cave. Second World War-era construction shored up the path leading to the cave and lowered the floor of the cave to aid visitor access. 'In 1947 alone, they dug out 600 cubic meters of sediment to make an entrance and concrete path and installed lighting for the public,' Jean Clottes points out. This was equivalent to removing eight 12m (39ft) shipping containers of sediment in order to pave the way for tourists. Although archaeologists and Palaeolithic art specialists have studied the paintings at Lascaux, providing dates and some context to the discovery, the cave was turned into a tourist site and conceptualised as an important part of the *longue durée* of French history, reaching all the way back to the Pleistocene. Lascaux was opened to the public on 14 July 1948.

As early as 1955, however, researchers noticed that some of the Palaeolithic art at Lascaux was beginning to deteriorate. Mould, fungi and bacteria had started to grow on the walls, obscuring some of the paintings and eating away the pigments from others. They traced this horrifying set of burgeoning growths to the carbon dioxide breathed out by the hordes of Lascaux's adoring visitors – sometimes as many as 1,000 per day, year after year. This increase of carbon dioxide, combined with visitors' body heat, warmed the cave and inadvertently offered an ecosystem for things to grow. Moreover, as visitors exhaled their acidified water vapour, their breath corroded away the rock faces, which in turn carried away pigment from the cave walls' surfaces. Essentially, Lascaux had been turned into a veritable petri dish.

Artificial ventilation was installed in 1958, the temperature was fixed at 14°C (57°F) and additional electric lighting was connected for use during established visiting hours as conservators began to battle the cave's problematic budding biota. The stubborn algae ignored these efforts to safeguard the paintings, and within a few years green patches of *Bracteacoccus minor* were growing along the cave's walls. Lascaux's conservators repeatedly treated the unsightly and destructive splotches with formaldehyde before discovering that this was doing more harm than good.

Lascaux closed in 1963 in an effort to combat the mould, fungus, lichens and such that continue to plague the cave today. Additional environmental controls were installed in 1966, and by 1979 the cave's climatic environment was considered stable. But by March 2000, conservators were facing mats of bacterial and fungal colonies. In the summer of 2001, they found white fungal growth on the floor and the backs of the walls of the Hall of Bulls. By May 2012, a new species of fungus, *Ochroconis lascauxensis*, had been found in the cave and named after Lascaux. ('Life finds a way.')

In addition to the constant biotic growths, the complexity of further conservation efforts at Lascaux has been compounded. When Lascaux's air-conditioning and percolating water-recovery systems were replaced in early

2001, for example, amid intensive rain, water pooled while the entrance to the cave was open and the systems were shut off for their replacement. Workers reported 'significant thermic and hygrometric disturbances' as they worked to combat these logistic issues. All of these efforts were to mitigate damage from previous decades. By 2007, UNESCO had threatened to place Lascaux on its World Heritage in Danger list.

Today, conservators have limited the total number of hours of human contact – that is, human presence – to 800 hours per year in the cave, and those 800 hours have to account for maintenance as well as academic research. Visitors wear sterile white coveralls, hairnets, gloves and booties over their feet. (Earlier conservation measures had actually required visitors to dip their toes in fungicide, but this caused too many problems as the fungicide destabilised the floor of the cave.) The entrance to the cave is guarded by two airlocks as conservators at Lascaux battle the Sisyphean task of trying to stave off further damage to the paintings and the cave. It goes without saying that tourists simply do not visit Lascaux any more, and haven't for decades.

But simply closing the site to the public didn't stop the destruction of the cave; the effects of hundreds of thousands of visitors in the twentieth century meant that Lascaux had, in effect, been spent. We cannot get it back.

Two hundred metres from Lascaux, however, a replica of the cave receives some 300,000 tourists a year. This replica was built in 1983 as a small-scale copy of the original cave, and an alternative way for tourists to 'see' the cave's stunning and iconic art. It quickly came to serve as a tourist and educational stand-in for a famous archaeological site. Known as Lascaux II – or sometimes derisively as the 'faux Lascaux' – it was hardly the first replica of a Palaeolithic site to open to tourists.

Altamira Cave, a Paleolithic site in the Cantabria region of Spain, underwent a similar story of discovery and tourism overuse. Discovered in 1879, Altamira closed to tourists in 1977 and reopened with limited access in 1982. The first replica of Altamira was built in the 1960s; a copy of the cave's painted ceilings was installed in the Deutsches Museum in Munich. This replica established methods of creating silicone moulds of the Palaeolithic cave, utilising the then-new stereo-scanning technology. A subsequent replica of Altamira was built in 2001 in Spain. But Lascaux II forcibly solidified the concept that building a replica was a good and reasonable way to negotiate balancing tourism and research access with preservation, particularly on site.

Today, there are three replicas of Lascaux: the 'original' copy of the cave from 1983, and an even 'more authentic' replica, known as Lascaux IV, built in 2012–2016. 'Inside the cave facsimile, the atmosphere is damp and dark, re-creating the humidity within the caves. Sounds are muffled; the temperature drops to about 16 degrees Celsius,' Lascaux IV's website boasts. 'This sequence is dedicated to contemplation, allowing people an experience of the sanctuary that once was. Lights flicker just as the animal fat lamps of Paleolithic times did, revealing the layers of paintings and engravings on the surface of the walls.' The third replica – Lascaux III – was built in 2012 as a mobile travelling set of replica panels that can move from exhibition to exhibition, bringing Lascaux to a variety of museum visitors.

Just as the destruction of Lascaux served as a cautionary tale for the preservation of newly discovered Palaeolithic cave art, building a replica has begun to be an accepted – perhaps even expected – method of providing visitors the opportunity to 'experience' and 'see' a site without tourist-ing it to death. In a bit of anthropogenic, modern irony, however, vehicle exhausts as a result of visitors parking their cars at

Lascaux II have added to the issues of conservation for both the original and the replica of the site.

★ ★ ★

With all these lessons to be learned from history, opening Chauvet to visitors was simply out of the question – and, to be honest, wasn't really ever an option that was on the table. (Recall that one of the first things conservationists did upon the discovery of Chauvet Cave was to physically barricade it from visitors and even monitor how much monitoring the cave could tolerate.) 'One absolute requirement is to make sure that the cave, its walls, climate and floors are preserved,' Jean Clottes explains in *Return to Chauvet Cave*. 'We must leave our successors an intact cave in which all kinds of research are still possible.'

In 2014, almost 20 years after its discovery, UNESCO added Chauvet Cave to its World Heritage List, thus cementing Chauvet's status as an icon of Upper Palaeolithic art and an important piece of humanity's deep cultural and evolutionary history. 'It is a welcome official recognition of the outstanding importance of the Cave in particular and of cave art in general. For preservation reasons, the cave will never be opened to the public,' French Minister of Culture Aurelie Filippetti declared at the press conference celebrating Chauvet's addition to the World Heritage List. 'An ambitious replica (called Caverne du Pont d'Arc – Ardèche) is however being built … It will allow all those interested to visit the reconstructed main panels as if one were inside the cave.'

This isn't just political glad-handing. The statement is a significant transition about what counts as 'the cave'. It begs the question of what, exactly, people are talking about when they talk about Chauvet. Certainly, the UNESCO listing applies to the tangible limestone *grotte* that Chauvet,

Brunel Deschamps and Hillaire discovered, affirming that
the Palaeolithic art and the cave that it's in has the same
cultural importance as the Vatican City or the Taj Mahal.
But unlike many of the other sites on the UNESCO list,
this protected World Heritage Site cannot – ought not –
be seen.

To that end, Filippetti's statement is also an acknowledge-
ment that what we mean when we talk about 'Chauvet
Cave' extends well beyond just the physical cave and
its paintings. For twenty-first-century audiences, to talk
about Chauvet Cave in terms of heritage is now also to
implicitly refer to the cave's replica, the Caverne du Pont
d'Arc, as this is how audiences see and experience the
cave's art since they can't see the originals for themselves.
To date, it's the only replica of a UNESCO-recognised
World Heritage Site.

So what does it mean to create a copy of an archaeological
site? What makes a copy good or bad? Can a replica really
stand in for the original thing, or does it have its own
cultural cachet? And do these copies of archaeological sites
themselves evolve over time?

Speaking about Chauvet Cave specifically, British
archaeologist Nicholas James ponders if such replicas of
Palaeolithic paintings can be authentic enough 'so as to
let visitors recognize, understand or experience something
of the ancient way of life and thought?' In an op-ed for
archaeology's leading journal, *Antiquity*, he pushes the
question even further. 'Does replication distract or detract
from the original? However faithful to the original, a
replica is "heritage", not an archaeological resource.'

Replicas of any archaeological site pose both technical
and ethical questions. First, building a replica – a good
replica, that is – isn't just slapping together a diorama and
corralling tourists through it. For all intents and purposes, an
archaeological   replica   involves   creating   an   exact

three-dimensional copy at full scale. This concept is hardly new. For close to 200 years, classical archaeologists have made casts of important artefacts, allowing copies to help facilitate professional study and augment museum collections. (Numerous Roman statues that have been cast for modern collections were in fact themselves copies of earlier ancient Greek works, making it even less clear just where and how to separate off an original from its copy.) Today, many museums opt to display replicas of artefacts for a plethora of reasons – an original artefact may be considered too culturally sensitive to display to audiences, for example, or including replicas can help augment objects that are on display. Making a copy of an entire archaeological site pushes that question even further and is a highly complex project that draws on expertise from engineers, artists and archaeologists.

Planning for the replica of Chauvet Cave began in 2007, when the Rhône-Alpes Region and the Department of Ardèche – with the support of the French State and the European Union – partnered with public and private funders. In 2013–2015, 500 artists, engineers, architects and special-effects designers built a copy of Chauvet based on 3D models created from 700 hours of laser scanning of the original cave. It took five years of research and assessment to model the replica, followed by a solid 30 months of construction, involving 35 different companies.

To begin with, the panels for the cave were made in the towns of Montignac and Toulouse, by Arc et Os as well as both the artist Gilles Tosello and the Guy Perazio firm. The teams used rock-coloured landscape mortar and resin, moulded over robust metal scaffolding, as the basis for the panels. In all, the replica used 130km (81mi) of handmade metal rods and 14,000 hangers for fixing and stabilising all of the panels. (This was a change from the design and engineering of earlier replicas. Engineers at Lascaux II used pneumatically placed concrete, and the Altamira ceiling panels made in the 1960s were resin. Moreover, suspending the replica of the Altamira ceiling at the Deutsches Museum was no easy task, as engineers had to work with the already built building's specs.) For the replica of Chauvet, the company, Deco Diffusion installed the panels on site and sculptors place-integrated the painted panels amid the rest of the replica. Photos of the installation show artists and technicians in white, all-covering Tyvek suits, trowelling the mortar-resin compound on to the replica walls and integrating the art with the replica's architecture. Based on the artists' outerwear and tools, I had to look at the pictures twice to check if the photos were of people excavating the original archaeological site or building its replica.

The panels represented 52 different types of rock – the decorated panels, of course, but also the 'blank' parts of the cave as well as the geological elements around the paintings. After all, Chauvet Cave isn't just the paintings. It's like a modern sculpture or art installation, where a work's composition and space are just as much part of the object as the obviously identifiable art.

The French press was permitted to see parts of the replica as the installation went up. For example, on 17 December 2014 – almost 20 years to the day after the original's discovery – the press was invited to see the 'Lion Panel' in the replica. (The contemporary 'Lion Panel' was actually drawn on polystyrene, which is a synthetic resin.) Ultimately, Chauvet's reproductions were vetted by the International Scientific Committee, a committee chaired by Jean Clottes and including Jean-Michel Geneste (prehistorian and archaeologist), Jean-Jacques Delannoy (geomorphologist) and Philippe Fosse (palaeontologist), all scientists who had studied the cave in the mid-1990s. Linkage to such expertise is a way to authenticate a replica as legitimate – Jean Clottes, as well as others, had thus authenticated both the original and its copy.

The air inside the replica is cool and moist, and the temperature sits at around 11°C (52°F) to offer a comparable environment to Chauvet Cave. 'It feels, and even smells, like a journey into a deep hole in the earth,' Joshua Hammer reported in *Smithsonian* magazine in 2015. 'But this excursion is actually taking place in a giant concrete shed set in the pine-forested hills of the Ardèche Gorge in southern France.' The Chauvet replica is two and a half times smaller than the original cave and cost €55 million.

The replica opened to the public on 25 April 2015 and was called the Caverne du Pont d'Arc. In its first two years it attracted more than a million visitors from 90 countries. More than 50 per cent of these people travelled to the Ardèche region for the sole purpose of seeing the replica. (The Caverne is about a kilometre from the original cave, a short drive from the village of Vallon.) Almost 30 per cent had come to the Ardèche for the first time, and 45 per cent had not been to the region for more than five years.

On average, visitors now spend three hours at the site, and the Caverne du Pont d'Arc has had over 24,000 guided tours. This has boosted tourism in the region, generating €25 million since the Caverne opened. Fabrice Tareau, director of the Caverne du Pont d'Arc, points out that in addition to that regional income, €100,000 – part of the royalties paid to the Syndicat Mixte and the Ardèche Department – was invested in the preservation of otherwise unprotected rural Ardèche heritage. Visitor tours rely on knowledgeable guides to take them through the replica cave and to interpret the Palaeolithic, helping to explain how archaeologists know what they do.

In addition to booking tours and trip planning, the Caverne du Pont d'Arc website contains a link to an impressive online model of Chauvet Cave, for web-based tourists to meander through while at home. Virtual visitors travel along the cave's metal walkways – with helpful arrows directing them from one room to another – and 360 degrees of images, stitched together. With people included for scale, the model is so much more real and understandable than simply seeing photographs of the cave in books. The experience is rather like Google Streetview. But instead of tooling through a neighbourhood, web-based visitors tour Chauvet Cave and are able to zoom in on clusters of bones, look up and see the cave's stalactites and, thanks to a small navigational map superimposed on the model of Chauvet Cave, see just where they are relative to the rest of the cave. It's almost as if the Caverne du Pont d'Arc is inviting visitors to verify for themselves that the physical replica they see is a faithful model of the original.

The Caverne du Pont d'Arc isn't only the recreated art panels or the replica of Chauvet – it also has an additional Aurignacian Gallery, school workshops, the 'La Terrasse' restaurant and the centre's shop, all a few minutes' walk from each other and all integrated into the landscape over the site's 12ha (30 acres). The centre has partnered with other Palaeolithic caves that are open to visitors, so Caverne tourists can see more of the archaeology of the Ardèche region. This model is not unlike sites in the Cantabria region of Spain such as the Tito Bustillo Cave, which use replicas as a bridge between parts of the cave that are open to visitors and parts that would be too environmentally sensitive to view.

The Caverne's Aurignacian Gallery is billed as a 'discovery centre', a place where visitors can learn more about the environment and lifestyle of humans living 36,000 years ago. The gallery has over 100 'experimental' archaeological artefacts, a theatre that seats 65 people and five life-sized Ice Age human models. The school workshop spaces offer cave-painting workshops to schoolchildren, and photos show casts of hominin ancestors from Africa that predate the

Ardèche regional story by millions of years. There's also story-time. ('How did prehistoric people light their fires? Younger children can discover the answer by listening to the story of the brave and curious Naly, who goes looking for this mysterious secret. Scenography and projected silhouette figures combine with the storyteller's words to create a timeless tale of discovery.') In 2017, the Caverne added a mock outdoor hunting area to its currently developing Palaeolithic Camp, boasting a new shaded canopy to make the experience more pleasant in the hot summer months. It even hosted the European Science Festival Spear-throwing Championships that autumn.

'There's something very deeply set, I think, in humans that makes us seek out the authentic, that direct physical connection to past people,' Palaeolithic archaeologist Natasha Reynolds suggested when I interviewed her about the cachet of replicas of Palaeolithic sites. 'But there's plenty of genuine worth to the replicas. If you can suspend your disbelief, you can gain a lot of understanding from them that you can't get in any other way.'

To go to the Caverne du Pont d'Arc isn't just to go and see the art – it's to participate in all of the trappings of France's Palaeolithic heritage.

★ ★ ★

This brings us to some tricky ethical questions about archaeological replicas. What makes a replica good? Can a replica really be authentic? What responsibilities – if any – do replicas have to an original archaeological site? Can a replica ever really, truly stand in for the genuine thing?

Ever since its opening, the Caverne du Pont d'Arc has billed itself as the new standard for archaeological replicas for the Palaeolithic, especially with the re-replica-ing of Lascaux in 2016. The Caverne du Pont d'Arc also set an impressive set of expectations for what a replica of a site ought to look like and how it should be scientifically vetted.

But critics of the Caverne du Pont d'Arc and other archaeological replicas ask: isn't this just Disney-fying archaeology and history? Hosting spear-throwing contests? Advertising a site's restaurant with organically grown and locally sourced meals? Isn't this really, critics complain, akin to taking Plato's Allegory of the Cave and pretending that the shadows on the wall are real objects? To put it bluntly: where does the history stop and the pandering start?

These questions boil down to maintaining the integrity of an archaeological site, forcing audiences to think about the question of *why* a site ought to be replicated, not just on the technical problems of actually doing it. Is it more valuable to know that a resource is preserved? Or is it, as archaeologist Nicholas James asks, more important to know that perpetual preservation cannot be guaranteed and that visitors who enjoy seeing the 'real thing' do it while spending that very site that they're coming to view?

'It is a veritable museum,' Chauvet, Brunel Deschamps and Hillaire said of the original cave in *Dawn of Art*. 'Wherever one looks, one is gripped by the beauty of the mineral forms or the drawn figures.' Unlike the Metropolitan Museum of Art or the Louvre, however, Chauvet Cave is a museum where even the simple act of observing the paintings destroys them and, because the paintings' galleries are part of the cave's architecture, the art is impossible to move.

How to balance heritage tourism with conservation constitutes a never-ending act and is completely dependent on the archaeological site in question. In Virginia, for example, the Colonial Williamsburg Historic Area is an open-air museum of eighteenth-century colonial life, employing re-enactors, craftspeople, historians and curators to both provide historical research and reach tourists. Part of what makes Colonial Williamsburg authentic – and a point that Colonial Williamsburg emphasises – is the ongoing research, archaeology, education and restoration that are inherently part of the site. In Honduras, casts of Classic Maya stelae are at the site of Copán; as so many tourists visit Copán, certain stelae have been moved to the site museum for protection and preservation. Visitors see replica stelae, but they see them as part of the original site. While museums have long presented ethnographic rooms or street scenes, they're generally generic reconstructions. They're platonic 'types' of scenes that one could expect to see; however, there aren't any claims about one specific site's authenticity. They're real, but they're removed from their originals in several ways.

'It is no longer a question of imitation, nor of reduplication, nor even of parody,' French philosopher Jean Baudrillard argues as he tries to tease out just when a copy of an object – a simulacra – becomes its own real, original, authentic thing. 'It is rather a question of substituting signs of the real for the real itself … Never again will the real have to be produced: this is the vital function of the model in a system of death, or rather of anticipated resurrection which no longer leaves any chance even in the event of death.'

Back at the Caverne du Pont d'Arc, however, for all of the 3D scanning, modelling and cast making, there are significant differences between the original cave and its copy. The building's plan and size, for example, differ from Chauvet Cave's topography. (Recall that the replica is much smaller than the original cave.) Where Chauvet Cave is a one-way geologic feature, the Caverne allows for the traffic flow of visitors to work their way through. At the risk of sounding flippant, the original cave is not equipped with emergency exits or designed to facilitate hundreds of thousands of visitors, as is the Caverne.

While every panel is brilliantly copied, it's clear that exactly copying the cave isn't conducive to the goals of the Caverne. By changing the layout of the site, it treats the cave as a platonic entity, rather than as the singularity of culture and geography that it is. Where the original Chauvet Cave has been posited to be a site of shamanistic ritual (or at least of artistic reflection), the purpose of the Caverne is to tell visitors about the Palaeolithic, have them marvel at the art and then to move them along. These differences highlight the difficult trade-offs that makers of archaeological replicas, especially those that copy entire archaeological sites, have to reckon with.

Replicas, reproductions, simulacra and copies have long held a reputation for being nothing more than inauthentic knock-offs. The art critic Jonathan Jones said as much in his *Guardian* review of the Caverne du Pont-d'Arc. 'No art lover wants to see a replica Rembrandt, a fake Freud or a simulacra [sic] of Seurat,' he sniffed, his distain palpable. 'Why then is it considered perfectly reasonable to offer fake Ice Age art as a cultural attraction?' While acknowledging that yes, viewing Palaeolithic art could damage the cave panels, Jones simultaneously refuses to see the Caverne du Pont-d'Arc as anything but a faux Chauvet. The replica, Jones argued, did nothing to really connect the viewer with the original art. (He generously allowed, however, that Werner Herzog's 'beautiful film', *Cave of Forgotten Dreams*, could be an acceptable alternative to experiencing the authenticity of Chauvet, a claim that moves beyond irony as one is still looking at pictures of the real thing and not the cave paintings themselves.)

But this assessment of archaeological replicas assumes that the only point of a replica is to inelegantly mimic the original work of art. It also assumes that the 'art' of the replica can't, itself, evolve and change. The Caverne du Pont d'Arc takes aim at these assumptions.

Both of these claims are at odds with the cultural cachet that replicas are beginning to accrue. It's clear that the sensory experience of Palaeolithic art underscores the aesthetic and engineering decisions about the Caverne du Pont-d'Arc even more so than those of Lascaux – that it's trying to recreate as much of the original cave as possible. A replica of Altamira Cave in Spain has had good tourism success, despite battling many of the same concerns about replicas and authenticity that front Chauvet and Lascaux.

There are numerous other reproductions of smaller and less iconic caves. ('Following the lead of France and Spain, rock art can be meticulously reproduced in settings that are open to the public, affording better protection for the actual sites,' Jean Clottes argues in his scholarly publications on the topic. 'Local communities must be afforded an economic stake in any programs to preserve rock art or increase cultural tourism.') And beyond Palaeolithic caves, sites like Tutankhamen's tomb have introduced replicas to reduce the wear and tear on original sites. More and more museums are offering virtual reality tours to their patrons, as cost-effective, visually stunning ways to 'see' a site without having to physically recreate it.

Dismissing replicas completely assumes that all art and artefacts can and ought to be viewed the same way, regardless of their material and context – it presumes that you should look at a panel of lions at Chauvet Cave in the same way that you look at the *Mona Lisa*. But this simply isn't the case. 'With Paleolithic cave paintings, the art is fundamentally, physically part of the gallery,' Palaeolithic archaeologist Aitor Ruiz-Redondo explained to me. 'We can't just create an environment to put around it to preserve it. This art, unlike Paleolithic artifacts that are portable, is in its own environment.'

When Jonathan Jones recounts his adolescent disappointment at Lascaux's replica, he suggests that deception is forever at the essence of a replica. ('What a farce, to promise cave art and deliver only a simulation.') But such an argument assumes that replicas themselves are static things, forever stuck with a mindset that they can only ever be poor approximations. This is demonstrably untrue. But it does require replica makers to be upfront about what a replica can and cannot do and be, and for tourists to be aware of this while visiting. 'Rather than thinking of replicas as knock-offs,' philosopher Erich Hatala Matthes proposes in an *Apollo* magazine essay, 'we could conceive of them as akin to maps or models. They offer us a vantage point that is often otherwise unavailable.'

Caverne du Pont-d'Arc is generations removed from the first Lascaux and Altamira replicas, showing that the art and engineering of copies is quietly and successfully claiming artistic space of its own. Lascaux IV, for example, used methods and models from the Caverne, as the Chauvet Cave replica had already shaped visitor expectations about the technological and aesthetic evolution of Palaeolithic replicas – no one would confuse Lascaux II with Lascaux IV. To that end, the differences between Lascaux II and Lascaux IV illustrate the evolution and cultural cachet that replica sites carry.

Interestingly, however, although Lascaux IV has evolved technologically, and modelled itself after the Caverne in terms of facilitating visitors' experiences and interpretation of Palaeolithic art, it has failed to attract visitors in the numbers that it had hoped. 'The world's most famous archaeological replica, Lascaux II was replaced in December 2016 by Lascaux IV. IV deserves to inherit the replication, but it is already struggling to cope as it seeks to outdo II's tally of visitors,' Nicholas James points out in his *Antiquity* articles. 'The ironies are manifold.'

★ ★ ★

Consequently, the Caverne du Pont-d'Arc is more than just a replica of Chauvet Cave – it is the Caverne du Pont-d'Arc. It offers something the original can't, an opportunity to see and experience art. (The art critic and philosopher Walter Benjamin claimed that a work 'becomes "authentic" only after the first copy ... is produced.') It is a good, authentic replica because it is completely honest about the trade-offs it is making between conservation and access.

On 2 April 2018, *The Art Newspaper* reported that a 'historic' legal compromise had recently been reached – one that guaranteed the three discoverers of the famous Palaeolithic Chauvet Cave in southern France a cash settlement as well as a percentage of tourist-charged admission. The French government had already paid the Chauvet Cave discoverers roughly €137,000 apiece as a reward for their discovery, but this recent financial ruling wasn't about rights to the original Chauvet Cave. This was about financial compensation regarding the Caverne du Pont d'Arc.

The three speleologists have an important tie to the modern history of Chauvet Cave, and it's this historical connection that they are receiving financial and legal compensation for. In order to fold the story of Chauvet's discovery into the Caverne du Pont-d'Arc, 'the association of the Caverne du Pont-d'Arc will now pay the three speleologists €50,000 for the image rights and the Chauvet name,' *The Art Newspaper* noted, 'and they will receive 1.7% of the admission fees to the replica cave.'

Awarding financial compensation and image rights to the Chauvet Cave discoverers (all of whom are and have been extremely invested in the region's palaeo-heritage) for the Caverne du Pont-d'Arc is a new step in the world of archaeological replicas – perhaps a part of the replicas' own cultural evolution – making the copy all the more Real.

As Seen In the British Museum

In May 2005, the British artist Banksy sauntered into the British Museum carrying a plastic bag and wearing a long overcoat and a fake beard. He ambled his way through various exhibits until he reached room 49, which houses the museum's Roman Britain collection. Checking to make sure no one was looking, he pulled out a piece of graffitied cement from the plastic bag and stuck it on the wall with strong adhesive tape, just below a small statue and to the left of several Roman-era figurines in a case. Labelled *'Peckham Rock'*, it looked at first glance like every other artefact in the early Britain collection.

Peckham Rock was accompanied by a label with a carefully mimicked museum caption, fake provenance and falsified index number. While the rock with its label could, maybe, pass for a genuine artefact, its iconography was a dead giveaway – even to the untrained eye. *Peckham Rock's* iconography featured a palaeo-inspired buffalo shot full of black arrows and a lumbering hominin-like figure pushing a shopping cart. Cultural geographer Luke Dickens described the figure as 'Neanderthal' and *Peckham Rock* as a conflation of 'British activist art' and 'cave painting'.

On the whole, *Peckham Rock* is a curious piece of art. It measures roughly 15 x 25cm (6 x 9in) and is made of shattered concrete, supposedly from Peckham in the London borough of Southwark. The museum label that Banksy installed with *Peckham Rock* describes the iconography as 'primitive art' from the 'Post-Catatonic era' with 'early man venturing towards the out-of-town hunting grounds'. The artist is credited as 'Banksymus Maximus', a prominent painter in the 'Post-Catatonic', and the label notes that the 'majority' of Banksymus Maximus's wall art has been 'destroyed by zealous municipal officials who fail to recognise the artistic merit and historical value of daubing on walls'.

This was hardly Banksy's first rogue museum installation. Two months earlier, Banksy had installed a 'harlequin beetle with airfix weapons' in the Hall of Biodiversity of New York's American Museum of Natural History. The specimen was labelled *Withus Oragainstus*, endemic to the United States, and purportedly lasted 12 days in the exhibition hall. 'Obviously, they've got their eye a lot more on things leaving than things going in, which works in my favour,' Banksy wryly noted in a 2005 interview with the BBC. Back at the British Museum, Banksy slipped away after installing the piece, leaving *Peckham Rock* to its fate. It didn't remain in an exhibition hall quite as long as the *Withus Oragainstus* beetle had, but *Peckham Rock* managed to last three days before it was found and promptly removed.

Peckham Rock challenges expectations about authenticity in a plethora of ways. The rock is actually from Hackney, not Peckham, to start with. It was carved out with a claw hammer and the motifs have an Upper Palaeolithic sort of vibe but are, obviously, contemporary. It's art that we might expect to be on the walls of Chauvet, Lascaux or even the Roman-built London Wall, Banksy seems to be arguing, if only our ancestors had been as absorbed with mass consumerism and capitalism as we are thousands of years later. Banksy ran a competition on his website for fans to take a photo of themselves with the artefact, offering a shopping cart as a prize.

Once the rock was unceremoniously removed from the wall of the British Museum, it was designated as 'lost property' and that was the end of it. Tom Hockenhull, a British Museum curator, said in a 2018 interview with the

Guardian, 'It was the cause of considerable embarrassment for the museum at the time and when Banksy asked for it back we were only too pleased to oblige.'

The Outside Institute, a short-lived gallery dedicated to graffiti, managed to borrow the piece for a collaborative exhibition in June 2005, at which *Peckham Rock* was shown as 'on loan' from the British Museum, further solidifying its genuine fakeness. 'We are proud to announce that we now have on display "the Peckham rock" kindly lent to us by the British Museum and the artist Banksy,' the label at Outside Institute read. 'This will not be displayed anywhere else and will be returned to the British Museum for historical verification at the end of the show. If you missed it in place at the British Museum we now offer you a chance to see it in a far grander environment.'

As fate would have it, Banksy's piece was to have another go at being on display in the British Museum. In August 2018, *Peckham Rock* was back, this time at the museum's request. It was one of a hundred objects that British radio host Ian Hislop gathered from the British Museum's extensive collections to display as part of an exhibition called *I Object!* The exhibition focused on ways that objects and artefacts can subvert cultural expectations and carry all the more cultural cachet for doing so. *Peckham Rock* has been the media's darling to build up interest and press around the exhibition's opening.

'At the time it was somewhat embarrassing but 13 years after Banksy installed a hoax exhibit at the British Museum curators are finally seeing the funny side,' the *Guardian* reported in August 2018. From its subversive arrival in the British Museum to becoming an official part of the museum's collection, *Peckham Rock*, it would seem, has a lot of life yet to live.

★ ★ ★

It's hard to imagine a better genuine fake than Banksy's *Peckham Rock*. It's subversive, it's complicated – over the course of its life it's been called a fake and a hoax and eventually it has become a legitimate artefact. More than anything else, *Peckham Rock* encourages audiences to examine the intent behind the piece. 'Put on display in one of the oldest museums in the world, the rock was intended to poke fun at the consumption habits of modern Britain, promote the work and name of an artist, assert the artistic and historical value of work of this type, and berate the purveyors and enforcers of "zero-tolerance" urban policy,' Luke Dickens neatly summarised in his review of the piece. *Peckham Rock* asks its audiences to decide if it's real and, more to the point, whether they think it's authentic. It's entirely possible that Banksy is offering us exactly what we don't know we want.

The question of intent fundamentally underlies everything about authenticity and fakes. Fakes with the intent to deceive are problematic – these sorts of fakes, to be blunt, are frauds. 'Although the objects involved in these frauds have been clearly shown to be forgeries, belief in them endures,' art historian Noah Charney outlines in *Art Forgery*. 'Regardless of the proof that they have always been fraudulent, many people refuse to believe that they are anything but authentic, and of huge importance. Even forgeries that are found out have the ongoing power to change history – just as the forgers had hoped.' How we internalise and respond to stories of fakes hinges on what we make of the fakers' intent to deceive their audiences.

These fakes are the frauds like the Spanish Forger's 'medieval' paintings being sold to unsuspecting tourists in Paris and ignorant collectors. 'Fossils' like Beringer's Lying Stones, created as a joke to put Dr Johann Bartholomew Adam Beringer of the University of Würzburg in his place. Natural diamonds stuffed into experiments to try to pass the lot off as laboratory-made. 'Maya codices' in which forgers can't even be bothered to match motifs to Mesoamerican art. William Henry Ireland's long-lost 'Shakespearean play', *Vortigern and Rowena*. Staging scenes in wildlife documentaries without telling audiences that the footage comes from a zoo and not the Arctic. These are all actual, technical, non-debatable examples of fakes as frauds – in each instance, the faker has traded authenticity for some sort of personal gain. As audience members who have effectively been pranked, we – as collectors, buyers, consumers or simply the naively unsceptical – are outraged at the deception and especially irked when we fall for it. 'By neglecting actual historical objects, and championing their reimagined counterparts, we efface the past,' historian Nir Shafir notes in an essay about fakes in the online magazine *Aeon*, pointing out that fake objects are often more 'believable' because they are, by design, intended to be more 'realistic' than their genuine counterparts.

On the other end of the spectrum, we have objects that are not easily dismissed as frauds – they have pristine provenances and unflappable origin stories. They include General Electric's laboratory diamonds, carefully grown under the watchful scrutiny of the company's team of scientists; flavours that, when run through a gas chroma-tograph, match the flavour notes of their agriculturally grown counterparts; the digital model of Chauvet Cave, and the cave's replica counterpart, the Caverne du Pont d'Arc. These things are all exactly what their makers have always claimed that they were.

The shift towards making authentic copies of objects has required an infusion of science and technology. For decades, science's role in the world of authentication was to ferret out fakes by evaluating the material make-up of pieces in question. Forensic tests became a way to offer proof about the legitimacy of a piece, based on whether the parts of the piece were from the time frame the provenance claimed. These tests became especially useful when expert dealers, collectors and scholars disagreed about a painting's authenticity. (Using chemical and radiometric tests to pinpoint an artefact's age also became increasingly popular in archaeology during this time.) Early forensic tests on paintings focused on examining the paint used; analyses developed later examined the pattern of paint cracking and any repetitive layers beneath a painting, as well as examining the painting's accompanying provenance documents, offering additional insights into other methods of authentication.

Today, as a result of tests like IR spectroscopy and microscopic analyses, infrared reflectography and a plethora of dating methods that rely on chemical isotopes – all methods that have been rigorously developed and tested within the scientific community – it would seem that it is harder and harder to pass off a fake to any savvy buyer or collector. Science's role grew, in the second half of the twentieth century, to not just sort out the less-than-legitimate fakes, but also to deconstructing the material element of a fake's natural counterparts. A diamond could not be grown in a laboratory before scientists knew what comprised a diamond and how the gem was made; flavours could not be synthesised on a mass scale until researchers could map grown-food flavour notes. Science and technology have offered ways for what we're calling genuine fakes to become much more authentic.

But many of these authentic objects are not perfect material matches for what we might call their 'real' counterparts. While they might be truthful objects, we're still left to grapple with what to make of them. And this is the space where we find that not all fakes are bad. Fakes that are made with the intent to conserve scarce or problematic resources might, in fact, be clever feats of science and engineering.

For example, as natural diamonds become more and more distasteful to consumers concerned about ethically sourced gems, laboratory-grown diamonds offer a way for a 'fake' to actually become more desirable than a natural diamond. Synthetic or artificial flavours offer a way for mass-produced foods to meet the needs of an ever-growing human population with food that is palatable. Archaeological sites are delicate, non-renewable pieces of global heritage; creating replicas of places like Chauvet and Lascaux offers a way to balance tourism with the needs of the archaeological sites. Rather than sniff at the distance between the original and its copy, it's time to look at the simulacrum as its own complete, authentic thing, possessing its own specific context and filling a unique set of ethical requirements.

Authenticity, as we see, is fluid, and over the course of their lives objects can be debunked and authenticated many times. To decide if a fake is good or bad or if a 'real' object is problematic or not requires nuance, finesse and an understanding of an object's historical context. Objects don't have an intrinsic morality to them – context is everything. It's not enough to simply refer to an object's material qualities – we need to refer to stories and histories that surround it.

Here is where the interaction between intent and audience expectations shifts. The Grolier Codex – so long dismissed as a fake because of its problematic provenance – offers such a narrative for objects moving backwards and forwards along authenticity's continuum. Collectors have eventually stopped seeing the Spanish Forger's works as cheap imitations of medieval prints; they became artefacts in and of themselves, indicative of the Forger's talent and their own bit of nineteenth-century art. Similarly, William Henry Ireland's letters – especially his forgeries of his original fakes – become collectable in their own right.

These are all ways that we've negotiated how to have real and fake things mash together to create genuine fakes.

★ ★ ★

I saw *Peckham Rock* as part of the British Museum's 2018 *I Object!* exhibition and it was fantastic. It was such a popular piece, in fact, that it took a while to wade through the throngs of other museum visitors who clustered around it.

Peckham Rock was, as promised, a piece of cement with a shopping cart and spikey-backed humanoid figure at the front and centre, surrounded by a stratigraphy of museum labels. The exhibit had Banksy's original faux museum label, of course, creased and looking a bit worse for wear, but there was a new label, off to the left, that explained the hoax and how the joke was on the British Museum. Visitors who weren't familiar with the story guffawed after reading that *Peckham Rock* had survived three days before the staff took it down ('Can you even imagine?'), and several voiced opinions that if they had seen it in room 49 they wouldn't have been taken in the way other, more gullible visitors were. ('It's not even very good,' one visitor announced to his wife. 'I would never think it's genuine.') But, mostly, visitors were charmed and amused at the audacity of a hoax that would challenge something as institutionally auspicious as the British Museum. On my way out of the exhibition, I bought a wooden copy of *Peckham Rock* for my office.

Fake has become a particularly charged word in the twenty-first century. To call something fake is no longer just about whether something is a fraud or phony. Fake has become a label, a judgement and a dismissal. However, if there is one thing to take away from the collection of stories in this book, it is this: we ought to be clear and deliberate about just what we are labelling, judging and dismissing. History, culture and context shape an object's authenticity, not decrees. And if we are to understand authenticity – the flip side of fake – then fake needs to be more than just a dog whistle.

Fakes need their stories, episodes, layered contexts, the 'I just can't believe it' instances, the dramatic reveals, the ever-evolving science of chasing down frauds and forgeries, and the moments when and if they are considered authenticated. The material life of any object isn't static, and the lives of genuine fakes are no different. When Mark Twain, in *The Innocents Abroad*, suggested the possibility that St Denis might have several skeletons if one were to reconstruct all of his remains, he nevertheless offered the observation that his mockery felt a little off. He was willing to acknowledge that maybe, even if all of the bones weren't necessarily real, the emotion that they and their reliquary elicited was real enough, and that he wasn't sure just how far to push his snark.

Genuine fakes offer an opportunity to explore how, why and under what circumstances we can – and ought – to accept things as authentic. Before we demand that something be authentic or dismiss something as fake, we ought to think about the purpose, intent and context of the object in question and what we would accept as the Real Thing. Those components matter because they mean that the status of these objects is ever-changing and ever-evolving.

Fakes may well let the world be deceived, as the Roman philosopher Petronius suggested, but that doesn't mean that there isn't important cultural history and meaning in our genuine fakes. The stories of their authenticity are still unfolding.

Bibliography

Introduction: Warhols Without Warhol

A matter of opinion – ARTnews. Accessed 11 January 2019. www. artnews.com/2012/02/28/a-matter-of-opinion.

Art authentication board – an idea that fell through. Widewalls. Accessed 11 January 2019. www.widewalls.ch/art-authentication-board.

Artist Paul Stephenson poses contentious questions of authorship with *After Warhol* series. It's Nice That, 26 October 2017. www.itsnicethat.com/news/paul-stephenson-after-warhol-prints-art-261017.

Baudrillard, Jean. 1994. *Simulacra and Simulation (The Body, in Theory)*. Ann Arbor, MI: University of Michigan Press.

———. 1996. *The System of Objects*. New York, NY: Verso.

Benjamin, Walter. 2008. *The Work of Art in the Age of Mechanical Reproduction*. Penguin Great Ideas 56. London: Penguin.

Charney, Noah. 2015. *The Art of Forgery: The Minds, Motives and Methods of the Master Forgers* (1st ed.). London; New York, NY: Phaidon Press.

———. Is there a place for fakery in art galleries and museums? Aeon Essays. *Aeon*. Accessed 18 September 2016. https://aeon.co/essays/is-there-a-place-for-fakery-in-art-galleries-and-museums.

Denis Dutton on art forgery. Accessed 14 August 2016. www.denisdutton.com/artistic_crimes.htm.

Dutton, Denis. Artistic crimes. *The British Journal of Aesthetics* 19 (1979): 302–41.

———. Authenticity in art. In *The Oxford Handbook of Aesthetics*, edited by Jerrold Levinson. 2003. New York, NY: Oxford University Press,.

Geurds, Alexander & Laura Van Broekhoven. 2013. *Creating Authenticity: Authentication Processes in Ethnographic Museums*. Leiden: Sidestone Press.

Hogenboom, Melissa. Can you love a fake piece of art? *BBC News*, 4 June 2012. www.bbc.com/news/magazine-18180057.

'It's like DJing with paintings': artist Paul Stephenson on being Warhol, 30 years after his death – I. Accessed 5 September

2018. https://inews.co.uk/culture/arts/like-djing-paintings-artist-paul-stephenson-recreating-warhol.

Lanchner, Carolyn. 2008. *Andy Warhol*. New York: Museum of Modern Art: distributed in the United States and Canada by DAP/Distributed Art Publishers.

Mattick, Paul. The Andy Warhol of philosophy and the philosophy of Andy Warhol. *Critical Inquiry* 24, no. 4 (1998): 965–87.

Perman, Stacy. This is bad news for people who spend millions on art. *Authentication in Art*, 24 September 2015, sec. *Fortune*.

Stokstad, Marilyn. 2008. *Art History* (3rd ed.). Upper Saddle River, NJ: Pearson Prentice Hall.

Twain, Mark. 2010. *The Innocents Abroad*. Wordsworth Classics Edition. Ware, Hertfordshire: Wordsworth Editions Ltd.

Youngs, Ian. The artist making 'new' Warhol paintings. *BBC News*, 18 October 2017, sec. Entertainment & Arts. www.bbc.co.uk/news/entertainment-arts-41634496.

Chapter 1: This Solemn Mockery

A look at Belle da Costa Greene – rare book collections @ Princeton, 18 March 2014. https://web.archive.org/web/20140318181732/http://blogs.princeton.edu/rarebooks/2010/08/a_look_at_belle_decosta_greene.html.

Antiques Roadshow. PBS. Accessed 11 January 2019. www.pbs.org/wgbh/roadshow/season/6/indianapolis-in/appraisals/spanish-forger-painting--200106T28.

Ardizzone, Heidi. 2007. *An Illuminated Life: Belle Da Costa Greene's Journey from Prejudice to Privilege* (1st ed.). New York, NY: W. W. Norton & Co.

Backhouse, Janet. The Spanish Forger. *The British Museum Quarterly* 33, no. 1/2 (1968): 65–71.

Barker, Nicolas. 2012. Introduction. In *The Stuart B. Schimmel Forgery Collection & Other Properties*, 4–6. London: Bonhams.

Charney, Noah. 2015. *The Art of Forgery: The Minds, Motives and Methods of the Master Forgers* (1st ed.). London; New York, NY: Phaidon Press.

————. Is there a place for fakery in art galleries and museums? *Aeon*. Accessed 18 September 2016. https://aeon.co/essays/is-there-a-place-for-fakery-in-art-galleries-and-museums.

————. This is your brain on knockoffs: the science of how we trick ourselves into not believing our eyes. *Salon*. Accessed 29

January 2017. www.salon.com/2017/01/29/this-is-your-brain-on-knockoffs-the-science-of-how-we-trick-ourselves-into-not-believing-our-eyes.

Durrieu, Paul, Pol de Limbourg & Jean Colombe. *Les très riches Heures de Jean de France, duc de Berry.* Paris, Plon–Nourrit, 1904. http://archive.org/details/gri_33125010357792.

Freeman, Arthur. William Henry Ireland's 'authentic original dorgeries': an overdue rediscovery. *Houghton Library Blog* (blog), 24 October 2012. https://blogs.harvard.edu/houghton/files/2012/08/Ireland.pdf.

————. The actual originals – the TLS. Accessed 6 April, 2018. www.the tls.co.uk/articles/private/the-actual-originals.

Greene, Belle da Costa. Letter to Charles Cunningham, Curator of Paintings, Museum of Fine Arts, Boston, MA, 27 September 1939. 2006 Expansion, B2, 03 Vault, Bay 073 Shelf A. Morgan Library.

————. Letter to Charles Cunningham, Curator of Paintings, Museum of Fine Arts, Boston, MA, 6 November 1939. 2006 Expansion, B2, 03 Vault, Bay 073 Shelf A. Morgan Library.

————. Letter to Charles Cunningham, Curator of Paintings, Museum of Fine Arts, Boston, MA, 15 October 1941. 2006 Expansion, B2, 03 Vault, Bay 073 Shelf A.

Hoover, John Neal. Stuart B. Schimmel: 16 May 1925–4 January 2013. *The Papers of the Bibliographical Society of America* 107, no. 2 (1 June, 2013): 142–45. https://doi.org/10.1086/680793.

Ireland, William Henry & Richard White. 1969. *Confessions of William-Henry Ireland, Containing the Particulars of His Fabrication of the Shakspeare Manuscripts.* A new ed. Burt Franklin Bibliography & Reference Series. Essays in Literature and Criticism 30. New York: B. Franklin.

Kramer, Hilton. Art: Morgan offers first-rate fakes. *New York Times*, 26 May 1978, sec. Archives. www.nytimes.com/1978/05/26/archives/art-morgan-offers-firstrate-fakes.html.

Kurz, Otto. 1948. *Fakes: A Handbook for Collectors and Students.* London: Faber and Faber.

Lacroix, Paul. 1870. *The Arts in the Middle Ages and the Renaissance* (1st ed.). London: Chapman & Hall.

Malone, Edmond. 1796. *An Inquiry into the Authenticity of Certain Miscellaneous Papers and Legal Instruments, Published Dec. 24, MDCCXCV. and Attributed to Shakspeare, Queen Elizabeth, and Henry, Earl of Southampton: Illustrated by Fac-Similes of the*

Genuine Hand-Writing of That Nobleman, and of Her Majesty; a New Fac-Simile of the Hand-Writing of Shakspeare, Never before Exhibited; and Other Authentic Documents: In a Letter Addressed to the Right Hon. James, Earl of Charlemont. London : Printed by H. Baldwin, for T. Cadell, Jun. [etc.]. http://archive.org/details/inquiryintoauthe00malo.

Pierce Card, Patricia. 2004. *The Great Shakespeare Fraud: The Strange, True Story of William-Henry Ireland*. Stroud: Sutton.

Price, T. D. & J. D. Burton. Provenience and provenance. In *An Introduction to Archaeological Chemistry*. 2011. New York, NY: Springer.

Stewart, Doug. 2010. *The Boy Who Would Be Shakespeare: A Tale of Forgery and Folly* (1st Da Capo Press ed.). Cambridge, MA: Da Capo Press.

Stokstad, Marilyn. *Art History* (3rd ed.). Upper Saddle River, NJ: Pearson Prentice Hall, 2008.

The Stuart B. Schimmel Forgery Collection with an Introduction by Nicolas Barker & Other Properties. 2012. London: Bonhams.

Thompson, Erin L. 2016. *Possession: The Curious History of Private Collectors from Antiquity to the Present*. New Haven, CT; London: Yale University Press.

———. Email interview with author. 19 March 2018.

Voelkle, William M. 1978. *The Spanish Forger*. New York: Pierpont Morgan Library.

———. 1987. *Spanish Forger: Master of Deception*. Milwaukee, WI: Haggerty Museum of Art, Marquette University.

———. The Spanish Forger: master of manuscript chicanery. In *The Revival of Medieval Illumination*, edited by Thomas Coomans & Jan De Maeyer: 207–27. 2007. Leuven: Leuven University Press.

Wellesley, Mary. Forged lives. Roundtable. *Lapham's Quarterly*. Accessed 16 December 2016. www.laphamsquarterly.org/roundtable/forged-lives.

Chapter 2: The Truth About Lying Stones

Baldwin, Stuart A. Educational palaeontological reproductions: the story of a unique small business. *Geology Today* 2, no. 6 (1986): 186–88.

Bennett, Jim. Museums and the history of science: practitioner postscript. *Isis* 96, no. 4 (2005): 602–8. https://doi.org/10.1086/498596.

Beringer, Johann Bartholomäus Adam, Georg Ludwig Hueber, Melvin E. Jahn & Daniel J. Woolf. 1963. *The Lying Stones of Dr. Johann Bartholomew Adam Beringer, Being His Lithographiæ Wirceburgensis.* Berkeley, CA: University of California Press.

Bowler, Peter J. & Morus, Iwan Rhys. 2005. *Making Modern Science: A Historical Survey.* Chicago, IL: University of Chicago Press.

Chambers, Paul. 2002. *Bones of Contention: The Archaeopteryx Scandals* (1st ed.). London: John Murray.

Dinosaurs with laser beams on their heads. Burke Museum, 27 May 2015. www.burkemuseum.org/press/dinosaurs-laser-beams-their-heads-0.

Gibson, Susannah. 2015. *Animal, Vegetable, Mineral?: How Eighteenth-Century Science Disrupted the NaturalOrder* (1st ed.). Oxford: Oxford University Press.

Gould, Stephen Jay. 2001. *The Lying Stones of Marrakech: Penultimate Reflections in Natural History.* California: Three Rivers Press.

Grene, Marjorie. 2004. *The Philosophy of Biology: An Episodic History.* Cambridge; New York, NY: Cambridge University Press.

Grimaldi, David A., Alexander Shedrinsky, Andrew Ross & Norbert S. Baer. Forgeries of fossils in 'amber': history, identification and case studies. *Curator* 37, no. 4 (1994): 251–74.

Herbert, Sandra. 2005. *Charles Darwin, Geologist.* Ithaca, NY: Cornell University Press.

Hochadel, Oliver. One skull and many headlines: the role of the press in the Steinau Hoax of 1911. *Centaurus* 58 (2016): 203–18.

Howlett, Eliza. Beringer's Lying Stones: casts & images? Personal communication, email 10 February 2017.

Lane, Meredith A. Roles of natural history collections. *Annals of the Missouri Botanical Garden* 83, no. 4 (1996): 536–45. https://doi.org/10.2307/2399994.

Mallatt, Jon M. Dr Beringer's fossils: a study in the evolution of scientific world view. *Annals of Science* 39, no. 4 (July 1982): 371–80.

Mayor, Adrienne. 2011. *The First Fossil Hunters: Dinosaurs, Mammoths, and Myth in Greek and Roman Times: With a New Introduction by the Author.* Princeton, NJ: Princeton University Press.

Palmer, Douglas. Fatal flaw fingers fake fossil fly. *New Scientist* (blog). Accessed 26 January 2017. www.newscientist.com/article/mg14018990-400-fatal-flaw-fingers-fake-fossil-fly.

Pickrell, John. How fake fossils pervert paleontology [excerpt]. *Scientific American*. Accessed 11 January 2019. www.scientificamerican.com/article/how-fake-fossils-pervert-paleontology-excerpt.

Powell, Philip. Letter between Philip Powell and British Musem. Letter, 22 November 1988.

Pyne, Lydia. 2016. *Seven Skeletons: The Evolution of the World's Most Famous Human Fossils*. New York, NY: Viking.

Rowe, Timothy, Richard A. Ketcham, Cambria Denison, Matthew Colbert, Xing Xu & Philip J. Currie. The Archaeoraptor forgery. *Nature* 410, no. 6828 (March 2001): 539–40. https://doi.org/10.1038/35069145.

Rudwick, Martin J. S. 2014. *Earth's Deep History: How It Was Discovered and Why It Matters*. Chicago, IL; London: University of Chicago Press.

Simons, Lewis. Archaeoraptor fossil trail. *National Geographic* 198, no. 4 (2000): 128–32.

Spencer, Frank. 1990. *Piltdown: A Scientific Forgery* (1st ed.). New York, NY: Oxford University Press.

Sundaram, Mark. The classical bedrock of fossil. Accessed 14 February 2017. www.alliterative.net/blog/2015/5/26/the-classical-bedrock-of-fossil.

Taylor, Paul. Beringer's iconoliths: palaeontological fraud in the early 18th century. *The Linnean* 20 (2004): 21–31.

Thompson, Erin L. 2016. *Possession: The Curious History of Private Collectors from Antiquity to the Present*. New Haven, CT; London: Yale University Press.

Weiner, J. S., Kenneth Page Oakley & Wilfrid E. Le Gros Clark. 1953. *The Solution of the Piltdown Problem*. London: British Museum (Natural History).

Chapter 3: Carbon Copy

Artificial production of real diamonds. *Mechanics' Magazine, Museum, Register, Journal, and Gazette* 206 (2 August 1828): 300–301.

Aykroyd, W. R. 1935. *Three Philosophers (Lavoisier, Priestley and Cavendish)*. London: W. Heinemann Ltd.

Bergstein, Rachelle. 2016. *Brilliance and Fire: A Biography of Diamonds* (1st ed.). New York, NY: Harper, an imprint of HarperCollinsPublishers.

————.What the diamond industry's new campaign, 'real is rare,' says about marketing luxury to millennials. *Forbes*. Accessed 11 June, 2018. www.forbes.com/sites/rachellebergstein/2016/10/19/what-real-is-rare-the-diamond-industrys-new-campaign-says-about-marketing-luxury-to-millennials.

Bol, Marjolijn. Coloring topazes, crystals and moonstones: the making and meaning of factitious gems, 300–1500. In *F for Fakes: Hoaxes, Counterfeits and Deception in Early Modern Science*, edited by Marco Beretta & Maria Conforti, 108–29. 2014. Science History Publications. Leiden: Brill Publishers.

————. Topazes, emeralds, and crystal rubies. The faking and making of precious stones. *The Recipes Project* (blog). Accessed 31 May 2018. https://recipes.hypotheses.org/4659.

Caley, Earle Radcliffe & William B. Jensen. The Leyden and Stockholm Papyri: Greco-Egyptian chemical documents from the early 4th century AD. Oesper Collections in the History of Chemistry. Cincinnati, OH: University of Cincinnati, 2008.

Chinese made first use of diamond, 17 May, 2005. http://news.bbc.co.uk/2/hi/science/nature/4555235.stm.

Choi, Charles Q. I proposed with a synthetic diamond. 29 July 2016. *Popular Science*. Accessed 9 May 2018. www.popsci.com/i-proposed-with-diamond-grown-in-lab.

Conflict diamond, 20 October 2000. https://web.archive.org/web/20001020115731/http://www.un.org/peace/africa/Diamond.html.

Cremation diamonds. Eterneva. Accessed 24 May 2018. www.eterneva.com.

Donovan, Arthur. 1993. *Antoine Lavoisier: Science, Administration, and Revolution*. Blackwell Science Biographies. Oxford; Cambridge, MA: Blackwell.

Doughty, Oswald. 1963. *Early Diamond Days: The Opening of the Diamond Fields of South Africa*. London: Longmans.

Epstein, Edward Jay. Have you ever tried to sell a diamond? *The Atlantic*, February 1982. www.theatlantic.com/magazine/archive/1982/02/have-you-ever-tried-to-sell-a-diamond/304575.

————. *The Rise and Fall of Diamonds: The Shattering of a Brilliant Illusion*. 1982. New York, NY: Simon & Schuster.

Feinstein, Charles H. 2005. *An Economic History of South Africa: Conquest, Discrimination and Development*. New York, NY: Cambridge University Press.

General Electric diamonds. *Democrat and Chronicle*, 11 May 1955. www.newspapers.com/image/135557198/?terms=general+ electric+synthetic+diamond+Rochester.

H. Tracy Hall Foundation. Accessed 11 January 2019. www. htracyhall.org.

Hannay, James Ballantyne. On the artificial formation of the diamond. *Proc. R. Soc. Lond.* 30, no. 200–205 (1879): 450–61.

Hazen, Robert M. 1999. *The Diamond Makers*. New York, NY: Cambridge University Press.

Kaplan, Sarah. Forget the ring: lab-grown diamonds are a scientist's best friend. *Washington Post*, 13 February 2017, sec. Speaking of Science. www.washingtonpost.com/news/speaking-of-science/wp/2017/02/13/forget-the-ring-lab-grown-diamonds-are-a-scientists-best-friend.

Klein, Joanna. If diamonds are forever, your data could be, too. *New York Times*, 26 October 2016. www.nytimes.com/2016/10/27/science/diamonds-data-storage.html.

Kunz, George Frederick. 1938. *The Curious Lore of Precious Stones: Being a Description of Their Sentiments and Folk Lore, Superstitions, Symbolism, Mysticism, Use in Medicine, Protection, Prevention, Religion, and Divination, Crystal Gazing, Birthstones, Lucky Stones and Talismans, Astral, Zodical, and Planetary*. New York: Halcyon House.

Levy, Arthur V. 2003. *Diamonds and Conflict: Problems and Solutions*. Hauppauge, NY: Nova Publishers.

Nassau, Kurt. 1980. *Gems Made by Man* (1st ed.). Radnor, PA: Chilton Book.

New diamonds from General Electric. *Evening Independent*, 18 February 1955. www.newspapers.com/image/4044984/?terms=General+Electric+synthetic+diamond.

Ozar, Garrett. Making memorial diamonds. Phone interview, 8 June 2018.

Patterson, Scott & Alex MacDonald. De Beers tries to counter a growing threat: man-made diamonds. *Wall Street Journal*, 6 November 2016, sec. Business. www.wsj.com/articles/de-beers-tries-to-counter-a-growing-threat-man-made-diamonds-1478434763.

Phelan, Matthew. Synthetic diamonds lead Princeton team to quantum computing breakthrough. Inverse. Accessed 7 July 2018. www.inverse.com/article/46728-synthetic-diamonds-are-necessary-for-quantum-computing-privacy.

Pliny the Elder. n.d. *Natural History*. Vol. Book 33.

Poirier, Jean-Pierre. 1996. *Lavoisier, Chemist, Biologist, Economist*. Chemical Sciences in Society Series. Philadelphia, PA: University of Pennsylvania Press.

Resnick, Irven M., trans. 2010. *Albert the Great On the Causes of the Properties of the Elements: Liber De Causis Proprietatium Elementorum* (new ed.). Milwaukee, WI: Marquette University Press.

Revie, James. Heritage: the case of the Hannay diamonds. *New Scientist*, 21 February 1980, 591.

Revolutionary instruments: Lavoisier's tools as objets d'art. Science History Institute, 2 June 2016. www.sciencehistory.org/distillations/magazine/revolutionary-instruments-lavoisiers-tools-as-objets-dart.

Shigley, James E., ed. 2005. *Synthetic Diamonds*. Gems & Gemology in Review. Carlsbad, CA: Gemological Institute of America.

Shor, Russell. De Beers sees growing diamond demand. www.gia.edu/sites/Satellite?c=Page&cid=1495254118454&childpagename=GIA/Page/ArticleDetail&pagename=GIA/Wrapper&WRAPPERPAGE=GIA/Wrapper., n.d.

Sullivan, Paul. A battle over diamonds: made by nature or in a lab? *New York Times*, 9 February 2018, sec. Your Money. www.nytimes.com/2018/02/09/your-money/synthetic-diamond-jewelry.html.

Tennant, Smithson. On the nature of the diamond. *Philosophica Transactions of the Royal Society of London* 87 (1797): 123–27.

Tunnell, Christopher. Laboratory diamonds for engagement. Personal communication, email 13 June 2018.

Wellings, Simon. Some facets of the geology of diamonds. *Scientific American* Blog Network. Accessed 24 May 2018. https://blogs.scientificamerican.com/guest-blog/some-facets-of-the-geology-of-diamonds.

Wells, H. G. 'The Diamond Maker'. *The Pall Mall Budget*. 1894.

Why smart people buy cubic zirconia engagement rings. *Forbes*. Accessed 9 May 2018. www.forbes.com/sites/quora/2017/07/03/why-smart-people-buy-cubic-zirconia-engagement-rings.

Zwick, Edward. *Blood Diamond* film, 2006.

Chapter 4: A Fake of a Different Flavour

Agapakis, Christina. The essence of taste. *Scientific American* Blog Network. Accessed 11 January 2019. https://blogs. scientificamerican.com/oscillator/the-essence-of-taste.

Berenstein, Nadia. Flavor added: the sciences of flavor and the industrialization of taste in America. Unpublished PhD dissertation, University of Pennsylvania, 2017.

———. Designing flavors for mass consumption. *The Senses and Society* 13, no. 1, 2018: 19–40. https://doi.org/10.1080/17458 927.2018.1426249.

Bourdieu, Pierre. 1986. *Distinction* (1st ed.). London: Routledge.

Breslin, Paul. An evolutionary perspective on food and human taste. *Current Biology* 23, no. 9 2013: 409–18. https://doi. org/10.1016/j.cub.2013.04.010.

Broderick, James. The practical flavorist vs. the basic researcher. *Food Technology* 26 (1972): 37–42.

———. Reflections of a retired flavorist before he forgets: strawberry. *Perfumer & Flavorist* 17, no. 3 (1992): 33–34.

Buddies, Science. Super-tasting science: find out if you're a 'supertaster'! *Scientific American.* Accessed 11 January 2019. www.scientificamerican.com/article/super-tasting-science-find-out-if-youre-a-supertaster.

Chapman, Peter. 2007. *Bananas: How the United Fruit Company Shaped the World* (1st US ed.). Edinburgh; New York: Canongate.

Classen, Constance, David Howes & Anthony Synnott. Artificial flavours. In *The Taste Culture Reader: Experiencing Food and Drink,* Carolyn Korsmeyer. 2005. Oxford; New York, NY: Bloomsbury.

Company history, Jelly Belly Candy Company. Accessed 20 July 2018. www.jellybelly.com/company-history.

Flandrin, Jean-Louis & Massimo Montanari (eds). 1999. *Food: A Culinary History.* Trans. by Albert Sonnenfeld. New York, NY: Columbia University Press.

Holmes, Bob. 2017. *Flavor: The Science of Our Most Neglected Sense* (1st ed.). New York, NY: W. W. Norton & Company.

Hoover, Kara C. The geography of smell. *Cartographica: The International Journal for Geographic Information and Geovisualization,* 20 January 2010. https://doi.org/10.3138/ carto.44.4.237.

Hoover, Kara C., Jessie Roberts & J. Colette Berbesque. Market smells: olfactory detection and identification in the built environment. *BioRxiv*, 26 February 2018, 270744. https://doi.org/10.1101/270744.

Jelly Belly Candy Company, official website & online candy store. Accessed 20 July 2018. www.jellybelly.com.

Kavaler, Lucy. 1963. *The Artificial World Around Us*. New York: John Day Company.

Khatchadourian, Raffi. The taste makers. *New Yorker*, 16 November 2009. www.newyorker.com/magazine/2009/11/23/the-taste-makers.

Korsmeyer, Carolyn. 2002. *Making Sense of Taste: Food and Philosophy*. Ithaca, NY: Cornell University Press.

Laudan, Rachel. 2015. *Cuisine and Empire: Cooking in World History* (1st ed.). Berkeley, CA: University of California Press.

Laurent, Anna. 2016. *Botanical Art from the Golden Age of Scientific Discovery*. Chicago, IL: University of Chicago Press.

Lloyd, John Uri. 1883. *Pharmaceutical Preparations. Elixirs, Their History, Formulae, and Methods of Preparation … with a Résumé of Unofficinal Elixirs from the Days of Paracelsus*. Cincinnati, OH: R. Clarke & Company, http://archive.org/details/pharmaceuticalp00lloygoog.

Mayer, Johanna. Why doesn't fake banana flavor taste like real bananas? *Science Friday*, 27 September 2017. www.sciencefriday.com/articles/why-dont-banana-candies-taste-like-real-bananas.

Murphy, Kate. Not just another jelly bean. *New York Times*, 26 June 2008, sec. Small Business. www.nytimes.com/2008/06/26/business/smallbusiness/26sbiz.html.

Musicant, Ivan. 1990. *The Banana Wars: A History of United States Military Intervention in Latin America from the Spanish-American War to the Invasion of Panama*. New York, NY: Macmillan.

Ordonez, Nadia, Michael F. Seidl, Cees Waalwijk, André Drenth, Andrzej Kilian, Bart P. H. J. Thomma, Randy C. Ploetz & Gert H. J. Kema. Worse comes to worst: bananas and panama disease – when plant and pathogen clones meet. *PLOS Pathogens* 11, no. 11 (19 November 2015): e1005197. https://doi.org/10.1371/journal.ppat.1005197.

Patterson, Daniel & Mandy Aftel. 2017. *The Art of Flavor: Practices and Principles for Creating Delicious Food*. New York, NY: Riverhead Books.

Perry, Jana. Jelly Belly flavors. Interview with author. 22 August 2018.

Ruppel Shell, Ellen. Chemists whip up a tasty mess of artificial flavors. *Smithsonian Magazine*, May 1986, 78–88.

Shapin, Steven. Changing tastes: how foods tasted in the Early Modern Period and how they taste now. In *Salvia Smaskrifter*, vol. 14. Uppsala University, 2011.

Spackman, Christy. Perfumer, chemist, machine: gas chromatography and the industrial search to 'improve' flavor. *The Senses and Society* 13, no. 1 (2 January 2018): 41–59. https://doi.org/10.108 0/17458927.2018.1425210.

Spence, Charles. 2017. *Gastrophysics: The New Science of Eating*. New York, NY: Viking.

Speth, John. Putrid meat and fish in the Eurasian Middle and Upper Paleolithic: are we missing a key part of Neanderthal and modern human diet? *PaleoAnthropology* 2017, no. 44–72 (2017): 44–72.

Taste and smell: a new theory. *Scientific American*. Accessed 24 July 2018. https://doi.org/10.1038/scientificamerican04101869-234.

Weird and gross jelly bean flavors – JellyBelly.Com, Jelly Belly Candy Company. Accessed 20 July 2018. www.jellybelly. com/weird-wild-and-gross-jelly-beans/c/289.

Wrangham, Richard W. 2009. *Catching Fire: How Cooking Made Us Human*. New York, NY: Basic Books.

Yeomans, Martin R., Lucy Chambers, Heston Blumenthal & Anthony Blake. The role of expectancy in sensory and hedonic evaluation: the case of smoked salmon ice-cream. *Food Quality and Preference* 19, no. 6 (1 September 2008): 565–73. https://doi.org/10.1016/j.foodqual.2008.02.009.

Chapter 5: Taking a Look Through Walrus Cam

Alaska Department of Fish and Game. Alaska's game species. dfg. webmaster@alaska.gov.

———. Lemming suicide myth. Accessed 1 August 2017. www. adfg.alaska.gov/index.cfm?adfg=wildlifenews. view_article&articles_id=56.

Alaska Walrus Cam joins a menagerie of wildlife video streams. *NBC News*, 28 May 2015. www.nbcnews.com/science/ weird-science/walrus-cam-joins-menagerie-wildlife- video-streams-n366291.

BBC Worldwide Press Office – *The Blue Planet* set for movie release. Accessed 5 February 2018. www.bbc.co.uk/pressoffice/bbcworldwide/worldwidestories/pressreleases/2003/03_march/bp_movie.shtml.

Boswall, Jeffrey. Wildlife film ethics: time for screen disclaimers. *Image Technology* 80, no. 9 (1998): 10–11.

Bousé, Derek. 2000. *Wildlife Films*. Philadelphia, PA: University of Pennsylvania Press.

———. False-intimacy: close-ups and viewer involvement in wildlife films. *Visual Studies* 18, no. 2 (2003): 123–32.

Crowther, Bosley. The screen: Disney's 'Peter Pan' bows; full-length color cartoon, an adaptation of Barrie play, is feature at the Roxy. *New York Times*, 12 February 1953. www.nytimes.com/movie/review?res=940CE3DF1F3AE23BBC4A52DFB4668388649EDE&pagewanted=print.

Cruz, Robert. 2012. The animated roots of wildlife films: animals, people, animation and the origin of Walt Disney's *True-Life Adventures*. MA: Montana State University. https://scholarworks.montana.edu/xmlui/bitstream/handle/1/1127/CruzR0512.pdf?sequence=1.

Daston, Lorraine. On scientific observation. *Isis* 99, no. 1 (2008): 97–110. https://doi.org/10.1086/587535.

Daston, Lorraine & Katharine Park. 2001. *Wonders and the Order of Nature, 1150–1750* (revised ed.). New York, NY: Zone Books.

Daston, Lorraine & Peter Galison. 2010. *Objectivity*. New York, NY: Zone Books.

Davies, Gail. Networks of nature: stories of natural history filmmaking from the BBC. PhD, University College London, 1998. Explore.Org. https://explore.org.

Foote, Peter (ed.). 1996. *Olaus Magnus: A Description of the Northern Peoples, 1555, Vol. 1*. Trans. by Peter Fisher & Humphrey Higgens. London: Hakluyt Society.

Frozen Planet: controversial BBC climate change episode to air in America. *Telegraph*. Accessed 4 February 2018. www.telegraph.co.uk/news/earth/earthnews/8939592/Frozen-Planet-controversial-BBC-climate-change-episode-to-air-in-America.html.

Gingras, Murray K., Ian A. Armitage, S. George Pemberton & H. Edward Clifton. Pleistocene walrus herds in the Olympic Peninsula area: trace-fossil evidence of predation by hydraulic jetting. *PALAIOS* 22, no. 5 (2007): 539–45.

Gladdis, Keith. BBC's little white lie: Polar Bear cubs were filmed for *Frozen Planet* in a zoo, not the Arctic. *Mail Online*, December 12, 2011. www.dailymail.co.uk/news/article-2073024/BBCs-little-white-lie-Polar-bear-cubs-filmed-Frozen-Planet-zoo-Arctic.html.

Goldenberg, Suzanne. Extreme Arctic sea ice melt forces thousands of walruses ashore in Alaska. *Guardian*, 27 August 2015, sec. Environment. www.theguardian.com/environment/2015/aug/27/walruses-alaska-arctic-sea-ice-melt.

Horak, Jan-Christopher. Wildlife documentaries: from classical forms to reality TV. *Film History: An International Journal* 18, no. 4 (2006): 459–75.

Krupnik, Igor & G. Carleton Ray. Pacific walruses, indigenous hunters, and climate change: bridging scientific and indigenous knowledge. *Deep Sea Research Part II: Topical Studies in Oceanography*, Effects of Climate Variability on Sub-Arctic Marine Ecosystems, 54, no. 23 (1 November 2007): 2946–57. https://doi.org/10.1016/j.dsr2.2007.08.011.

Lambert, Laura. *Blue Planet II* is the biggest show of 2017: 14 million watch opener. *Mail Online*, November 7, 2017. www.dailymail.co.uk/~/article-5056457/index.html.

Lawrence, Natalie. Decoding the morse: the history of 16th-century narcoleptic walruses. *The Public Domain Review*. Accessed 15 June 2017. /2017/06/14/decoding-the-morse-the-history-of-16th-century-narcoleptic-walruses.

———. There be monsters: from cabinets of curiosity to demons within. *Aeon*. Accessed 12 January 2019. https://aeon.co/essays/there-be-monsters-from-cabinets-of-curiosity-to-demons-within.

Lendon, Brad. 35,000 walruses 'haul out' on Alaska Beach. CNN. Com. Accessed 31 May 2017. www.cnn.com/2014/10/01/us/alaska-massive-walrus-gathering/index.html.

Louson, Eleanor. Never before seen: spectacle, staging, and story in wildlife film's blue-chip renaissance. PhD, York University, 2018.

———. Taking spectacle seriously: wildlife film and the legacy of natural history display. *Science in Context* 31, no. 1 (March 2018): 15–38. https://doi.org/10.1017/S0269889718000030.

McIntosh, Steven. 22 things you need to know about *Blue Planet II*. *BBC News*, 29 October 2017, sec. Entertainment & Arts. www.bbc.com/news/entertainment-arts-41692370.

Minteer, Ben A. 2018. *The Fall of the Wild: Extinction, De-Extinction, and the Ethics of Conservation*. New York, NY: Columbia University Press.

Mitman, Gregg. 2009. *Reel Nature: America's Romance with Wildlife on Film* (2nd ed.). Seattle, WA: University of Washington Press.

Palmer, Chris. 2010. *Shooting in the Wild: An Insider's Account of Making Movies in the Animal Kingdom* (1st ed.). San Francisco, CA: Counterpoint,.

Planet Earth II's dangerous aestheticization of nature. *New Republic*, 14 February 2017. https://newrepublic.com/article/140252/view-kill-planet-earth-ii-review-bbc.

Richards, Morgan. The wildlife docusoap: a new ethical practice for wildlife documentary? *Television & New Media* XX, no. X (2012): 1–15.

———. Greening wildlife documentary. In *Environmental Conflict and the Media*, edited by Libby Lester & Brett Hutchins. 2013. New York, NY: Peter Lang.

Schuessler, Ryan. Indigenous cooperation a model for walrus conservation. *Hakai Magazine*. Accessed 12 January 2019. www.hakaimagazine.com/news/indigenous-cooperation-model-walrus-conservation.

Singh, Anita. *Frozen Planet*: BBC 'faked' Polar Bear birth. *Telegraph*. sec. Culture. 12 December 2011. www.telegraph.co.uk/culture/tvandradio/bbc/8950070/Frozen-Planet-BBC-faked-polar-bear-birth.html.

Sir David Attenborough forced to explain when animals are not filmed in the wild for new BBC documentary after fakery row. *Daily Mail Online*. Accessed 11 June 2017. www.dailymail.co.uk/news/article-2254131/Sir-David-Attenborough-forced-explain-animals-filmed-wild-new-BBC-documentary-fakery-row.html.

Sitka, Emily Russell, KCAW. In unnerving trend, 35,000 walrus haul out at Point Lay. *Alaska Public Media* (blog). Accessed 1 June 2017. www.alaskapublic.org/2015/09/11/in-unnerving-trend-35000-walrus-haul-out-at-point-lay.

Sohn, Emily. What now, walrus? *Hakai Magazine*. Accessed 19 February 2018. www.hakaimagazine.com/features/what-now-walrus.

Tanaka, Yoshihiro & Naoki Kohno. A new Late Miocene odobenid (Mammalia: Carnivora) from Hokkaido, Japan suggests rapid diversification of basal Miocene odobenids. *PLOS ONE* 10, no. 8 (5 August 2015): e0131856. https://doi.org/10.1371/journal.pone.0131856.

'The world has not improved:' David Attenborough on *Blue Planet II* and how the ocean needs our attention. *CBC News*, 15 October 2017. www.cbc.ca/news/entertainment/blue-planet-2-1.4350287.

Thomas, Bob. 1976. *Walt Disney: An American Original*. New York, NY: Simon and Schuster.

Weiss, Edward. P. & Ryan P. Morrill. Walrus Islands State Game Sanctuary Annual Management Report 2016. Division of Wildlife Conservation: Alaska Department of Fish and Game, 2016.

Yong, Ed. *Blue Planet II* is the greatest nature series of all time. *The Atlantic*, 16 January 2018. www.theatlantic.com/science/archive/2018/01/blue-planet-ii-is-the-greatest-nature-series-of-all-time/550583.

Chapter 6: The Great Blue Whale

Abbott, Sam. Whales smelled out the $$. *The Billboard*, 28 June 1952, 94–96.

About Blue Whales. Beaty Biodiversity Museum. Accessed 28 March 2017. http://beatymuseum.ubc.ca/whats-on/exhibitions/permanent-exhibitions/blue-whale-display/about-blue-whales.

Alberti, Samuel J. M. M. (ed.). 2011. *The Afterlives of Animals: A Museum Menagerie*. Charlottesville, VA: University of Virginia Press.

Blue Whale. *National Geographic*. Animals, 10 September 2010. www.nationalgeographic.com/animals/mammals/b/blue-whale.

Blue Whale skeleton 'Hope' takes centre stage in museum. Natural History Museum. Accessed 20 February 2018. www.nhm.ac.uk/press-office/press-releases/blue-whale-skeleton-takes-centre-stage-in-museum.html.

Burnett, D. Graham. 2012. *The Sounding of the Whale: Science and Cetaceans in the Twentieth Century* (1st ed.). London; Chicago, IL: University of Chicago Press.

Canterbury Museum's Blue Whale to return to public view. Staff. Accessed 26 March 2018. www.stuff.co.nz/environment/

83263626/canterbury-museums-blue-whale-to-return-to-public-view.

Chindahl, George L. 1959. *A History of the Circus in America.* Caldwell, ID: Caxton Printers.

Daston, Lorraine & Katharine Park. 2001. *Wonders and the Order of Nature, 1150–1750* (revised ed.). New York, NY: Zone Books.

de Roos, Michael. Big Blue. Email interview, 27 March 2018.

———. Re: Beaty museum exhibit?, Email interview, 28 March 2018.

Haraway, Donna. Teddy bear patriarchy: taxidermy in the Garden of Eden, New York City, 1908–1936. *Social Text*, no. 11 (1984): 20–64. https://doi.org/10.2307/466593.

Kohler, Robert. Finders, keepers: collecting sciences and collecting practice. *History of Science* 45, no. 4 (2007): 428–54.

Miller, Mark. *Raising Big Blue.* Documentary. Discovery Channel, 2011. https://vimeo.com/19403399.

Museum unveils 'Hope' the Blue Whale skeleton. Natural History Museum. Accessed 20 February 2018. www.nhm.ac.uk/about-us/news/2017/july/museum-unveils-hope-the-blue-whale-skeleton.html.

Nance, Susan. 2013. *Entertaining Elephants: Animal Agency and the Business of the American Circus.* Animals, History, Culture. Baltimore, MD: Johns Hopkins University Press.

Pfening, Fred. Moby Dick on rails. *The Bandwagon*, 1987, 14–17.

Poliquin, Rachel. 2012. *The Breathless Zoo: Taxidermy and the Cultures of Longing.* University Park, PA: Pennsylvania State University Press.

Pyenson, Nick. Your work with whales. Email interview, 21 March 2017.

———. 2018. *Spying on Whales: The Past, Present, and Future of Earth's Most Awesome Creatures* (1st ed.). New York, NY: Viking.

Raising Big Blue. Beaty Biodiversity Museum. Accessed 21 May 2016. http://beatymuseum.ubc.ca/whats-on/exhibitions/permanent-exhibits/blue-whale-display/raising-big-blue.

Rossi, Michael. Modeling the unknown: how to make a perfect whale. *Endeavour* 32, no. 2 (June 2008): 58–63. https://doi.org/10.1016/j.endeavour.2008.04.003.

———. Fabricating authenticity: modeling a whale at the American Museum of Natural History, 1906–1974. *Isis* 101, no. 2 (2010): 338–61. https://doi.org/10.1086/653096.

————. Whale models, history of science. Skype interview, 28 February 2017.

Scheer, Rachel. *Scientific American* co-hosts whale tweet-up at American Museum of Natural History. *Scientific American* Blog Network. Accessed 13 January 2019. https://blogs. scientificamerican.com/at-scientific-american/scientific-american-co-hosts-whale-tweet-up-at-american-museum-of-natural-history.

Stocking, George W. 1988. *Objects and Others: Essays on Museums and Material Culture.* Madison, WI: University of Wisconsin Press.

Stoddart, Helen. 2000. *Rings of Desire: Circus History and Representation.* Manchester: Manchester University Press.

UBC Blue Whale. Accessed 13 January 2019. www.cetacea.ca/ubc-blue-whale.html.

Van Gelder, Richard. Whale on my back. *Curator* XIII, no. 2 (1970): 95–118.

Yanni, Carla. 2006. *Nature's Museums: Victorian Science and the Architecture of Display* (1st ed.). New York, NY: Princeton Architectural Press.

Chapter 7: And Now It's The Real Deal

Arroyo, Barbara, Ronald L. Bishop, Oswaldo Chinchilla Mazariegos, John E. Clark, Barbara W. Fash, Virginia Fields, Stephen D. Houston *et al.* 2012. *Ancient Maya Art at Dumbarton Oaks.* Edited by Joanne Pillsbury, Miriam Doutriaux, Reiko Ishihara-Brito & Alexandre Tokovinine. Washington, DC: Dumbarton Oaks Research Library and Collection.

Art Authentication Board – an idea that fell through. Widewalls. Accessed 11 January 2019. www.widewalls.ch/art-authentication-board.

Batres, Leopoldo. *Antiguidades Mejicanas Falsificadas*, 1910. http://archive.org/details/BatresLeopoldoAntiguidadesMejicas FalsificadasCopy.compressed.

Bender, Rose. America's 'new' oldest book: researchers confirm the authenticity of the ancient Mayan Grolier Codex. *Yale Scientific Magazine* (blog), 11 January 2017. www.yalescientific. org/2017/01/americas-new-oldest-book-researchers-confirm-the-authenticity-of-the-ancient-mayan-grolier-codex.

Berger, Dina & Andrew Grant Wood. 2009. *Holiday in Mexico: Critical Reflections on Tourism and Tourist Encounters.* Durham, NC: Duke University Press.

Blakemore, Erin. New analysis shows disputed Maya 'Grolier Codex' is the real deal. Smithsonian. Accessed 13 January 2019. www.smithsonianmag.com/smart-news/maya-codex-once-thought-be-sketchy-real-thing-180960466.

Calvo Del Castillo, Helena, Ruvalcaba Sil, Jose Luis, Tomás Calderón, Marie Vander Meeren & Laura Sotelo. The Grolier Codex: A PIXE & RBS study of the possible Maya document, 2007. http://orbi.ulg.ac.be/handle/2268/111350.

Carter, Nicholas P. & Jeffrey Dobereiner. Multispectral imaging of an Early Classic Maya Codex fragment from Uaxactun, Guatemala. *Antiquity* 90, no. 351 (June 2016): 711–25. https://doi.org/10.15184/aqy.2016.90f.

Casas, Bartolomé de las & Manuel Serrano y Sanz. 1909. *Apologética historia de las Indias.* Madrid: Bailly Bailliére é hijos.

Charney, Noah. 2015. *The Art of Forgery: The Minds, Motives and Methods of the Master Forgers* (1st ed.). London; New York, NY: Phaidon Press.

Chinchilla Mazariegos, Oswaldo Fernando. 2017. *Art and Myth of the Ancient Maya.* New Haven, CT: Yale University Press.

Christenson, Allen J., trans. 2007. *Popol Vuh: The Sacred Book of the Maya: The Great Classic of Central American Spirituality, Translated from the Original Maya Text.* Norman, OK: University of Oklahoma Press.

Cline, Eric H. & Glynnis Fawkes. 2017. *Three Stones Make a Wall: The Story of Archaeology.* Princeton, NJ: Princeton University Press.

Coe, Michael D. 1973. *The Maya Scribe and His World.* New York, NY: The Grolier Club.

———. 1998. *Art of the Maya Scribe* (1st ed.). New York, NY: Harry N. Abrams.

———. 2012. *Breaking the Maya Code* (3rd ed.). New York, NY: Thames & Hudson.

Coe, Michael D. & Mark Van Stone. 2005. *Reading the Maya Glyphs* (2nd ed.). New York, NY: Thames & Hudson.

Coe, Michael D. & Stephen D. Houston. 2015. *The Maya* (9th ed.). New York, NY: Thames & Hudson.

Coe, Michael, Mary Miller, Stephen Houston, Karl Taube, Simon Martin, Takeshi Inomata, Daniela Triadan *et al.* 2015. *Maya Archaeology 3: Featuring the Grolier Codex.* San Francisco, CA: Precolumbia Mesoweb Press.

Cummings, Mike. Authenticating the oldest book in the Americas. YaleNews, 18 January 2017. https://news.yale.edu/2017/01/18/authenticating-oldest-book-americas.

FAMSI – Maya codices – the Grolier Codex. Accessed 3 August 2017. www.famsi.org/mayawriting/codices/grolier.html.

Geurds, Alexander & Laura Van Broekhoven. 2013. *Creating Authenticity: Authentication Processes in Ethnographic Museums.* Leiden: Sidestone Press.

Holmes, William Henry. 1882. *Pottery of the Ancient Pueblos.* Washington, DC: Smithsonian.

———. *Archaeological Studies Among the Ancient Cities of Mexico.* Anthropological Series. Field Columbian Museum; v. 1, No. 1. Chicago, 1895. https://catalog.hathitrust.org/Record/009443928. https://onlinebooks.library.upenn.edu/webbin/gutbook/lookup?num=41998.

———. *The Painter and the National Parks.* Washington: Govt. print. off., 1917. https://catalog.hathitrust.org/Record/009591420.

Kelker, Nancy L. & Karen Olsen Bruhns. 2010. *Faking Ancient Mesoamerica.* Walnut Creek, CA: Left Coast Press.

Kettunen, Harri & Christophe Helmke. 2014. *Introduction to Maya Hieroglyphs* (14th ed.). Comenius University in Bratislava: The Slovak Archaeological and Historical Institute.

Landa, Diego de. 1565. *Relación de las Cosas de Yucatán.*

Love, Bruce. Authenticity of the Grolier Codex remains in doubt. *Mexicon* 39, no. 4 (2017): 88–95.

Meyer, Karl E. 1973. *The Plundered Past.* New York, NY: Atheneum.

Milbrath, Susan. New questions about the authenticity of the Grolier Codex. *Latin American Indian Literatures Journal*, 18(1): 50–83. Accessed 3 August 2017.

Newitz, Annalee. Confirmed: mysterious ancient Maya book, Grolier Codex, is genuine. *Ars Technica*, 12 September 2016. https://arstechnica.com/science/2016/09/confirmed-mysterious-ancient-maya-book-grolier-codex-is-genuine.

Peterson, Daniel. Defending the faith: an ancient American book, dismissed as a fraud, proves to be genuine. DeseretNews.com, 22 September 2016. www.deseretnews.com/article/

865663001/An-ancient-American-book-dismissed-as-a-fraud-proves-to-be-genuine.html.

Ruvalcaba, Jose Luis, Sandra Zetina, Helena Calvo del Castillo, Elsa Arroyo, Eumelia Hernández, Marie Van der Meeren & Laura Sotelo. The Grolier Codex: a non destructive study of a possible Maya document using imaging and ion beam techniques. *MRS Online Proceedings Library Archive* 1047 (ed 2007). https://doi.org/10.1557/PROC-1047-Y06-07.

Sharer, Robert J. & Loa P. Traxler. 2006. *The Ancient Maya* (6th ed.). Stanford, CA: Stanford University Press.

Stephens, John Lloyd. 1993. *Incidents of Travel in Central America, Chiapas, and Yucatan.* Washington, DC: Smithsonian Institution Press.

Thompson, Erin L. Enduring appeal of fakes? Email, 19 March 2018.

Vail, Gabrielle. The Maya codices. *Annual Review of Anthropology* 35, no. 1 (2006): 497–519. https://doi.org/10.1146/annurev.anthro.35.081705.123324.

Vitelli, Karen D. The antiquities market. *Journal of Field Archaeology* 4, no. 4 (1 January 1977): 459–72. https://doi.org/10.1179/009346977791490168.

Yates, Donna. Museums, collectors, and value manipulation: tax fraud through donation of antiquities. *Journal of Financial Crime* 23, no. 1 (31 December 2015): 173–86. https://doi.org/10.1108/JFC-11-2014-0051.

———. Maya artefacts and Mexican bandits: trafficking tall tales. Accessed 14 August 2017. www.anonymousswisscollector.com/2014/07/maya-artefacts-and-mexican-bandits-trafficking-tall-tales.html.

———. Grolier Codex « trafficking culture. Accessed 16 August 2017. http://traffickingculture.org/encyclopedia/case-studies/grolier-codex.

———. Appeal of fake artifacts? Email interview, 19 March 2018.

Chapter 8: The Art of Making the Palaeolithic Come to Life

Bahn, Paul. Putting a brave face on a fake. Cambridge Core. *Cambridge Archaeological Journal* 6, no. 2 (1996): 309–10.

———. A lot of bull? Pablo Picasso and Ice Age cave art. *Munibe* 57 (2005): 217–23.

Bahn, Paul G. & Jean Vertut. 1997. *Journey Through the Ice Age.* Berkeley, CA: University of California Press.

Baudrillard, Jean. 1994. *Simulacra and Simulation (The Body, in Theory).* Ann Arbor, MI: University of Michigan Press.

————. 1996. *The System of Objects.* New York, NY: Verso.

Bradshaw Foundation. Chauvet Cave granted World Heritage status. Bradshaw Foundation. Accessed 25 June 2018. www.bradshawfoundation.com/chauvet/chauvet_cave_UNESCO_world_heritage_site.php.

Brown, Bill. Thing theory. *Critical Inquiry* 28, no. 1 (1 October 2001): 1–22. https://doi.org/10.1086/449030.

Caverne Du Pont d'Arc Press Kit. Caverne Pont d'Arc, 2017.

Chauvet, Jean-Marie, Eliette Brunel Deschamps & Christian Hillaire. 1996. *Dawn of Art: The Chauvet Cave: The Oldest Known Paintings in the World.* New York, NY: H. N. Abrams.

Clottes, Jean, ed. 2003. *Return to Chauvet Cave: Excavating the Birthplace of Art: The First Full Report.* London: Thames & Hudson.

————. Rock art: an endangered heritage worldwide. *Journal of Anthropological Research* 64, no. 1 (1 April 2008): 1–18. https://doi.org/10.3998/jar.0521004.0064.101.

Deutsches Museum. Altamira – Höhlenmalerei Der Steinzeit. Deutsches Museum, 2012.

————. Deutsches Museum: Altamira-Höhle. Accessed 13 January 2019. www.deutsches-museum.de/ausstellungen/kommunikation/altamira-hoehle.

————. Deutsches Museum: Literatur. Accessed 13 January 2019. www.deutsches-museum.de/sammlungen/meisterwerke/meisterwerke-vi/altamira-hoehle/literatur.

Furness, Hannah. Mary Beard: it doesn't really matter if tourists damage Pompeii. *Telegraph*, 6 April 2016. www.telegraph.co.uk/news/2016/04/06/mary-beard-it-doesnt-really-matter-if-tourists-damage-pompeii.

Hammer, Joshua. Finally, the beauty of France's Chauvet Cave makes its grand public debut. *Smithsonian Magazine.* Accessed 20 June 2018. www.smithsonianmag.com/history/france-chauvet-cave-makes-grand-debut-180954582.

Hatala Matthes, Erich. Digital replicas are not soulless – they help us engage with art. *Apollo Magazine*, 23 March 2017. www.apollo-magazine.com/digital-replicas-3d-printing-original-artworks.

————. Palmyra's ruins can rebuild our relationship with history – Erich Hatala Matthes. Aeon ideas. *Aeon*. Accessed 21 April 2018. https://aeon.co/ideas/palmyras-ruins-can-rebuild-our-relationship-with-history.

'Historic' agreement resolves dispute over Chauvet Cave (and replica). *The Art Newspaper*. Accessed 20 June 2018. www.theartnewspaper.com/news/historic-agreement-resolves-dispute-over-chauvet-cave-and-replica.

James, Nicholas. Replication for Chauvet Cave. *Antiquity* 90, no. 350 (April 2016): 519–24. https://doi.org/10.15184/aqy.2016.63.

————. Our fourth Lascaux. *Antiquity* 91, no. 359 (October 2017): 1367–74. https://doi.org/10.15184/aqy.2017.145.

Jones, Jonathan. Don't fall for a fake: the Chauvet Cave art replica is nonsense. *Guardian*, 15 April 2015, sec. Art and design. www.theguardian.com/artanddesign/jonathanjonesblog/2015/apr/15/chauvet-cave-art-replica-is-nonsense.

Korsmeyer, Carolyn. 2002. *Making Sense of Taste: Food and Philosophy*. Ithaca, NY: Cornell University Press.

————. Touch and the experience of the genuine. *The British Journal of Aesthetics* 52, no. 4 (1 October 2012): 365–77. https://doi.org/10.1093/aesthj/ays043.

————. 2019. *Things: In Touch with the Past*. Oxford; New York, NY: Oxford University Press.

Lascaux IV: The International Centre for Cave Art. Accessed 27 June 2018. /projects/322-lascaux-iv-the-international-centre-for-cave-art.

Lascaux's 18,000 year-old cave art under threat. Accessed 27 June 2018. https://phys.org/news/2011-06-lascaux-hands-off-approach-threatened-art.html.

Leadbeater, Chris. Lascaux Cave and the rise of the fake attraction. *Telegraph*, 5 February 2016. www.telegraph.co.uk/travel/destinations/europe/france/aquitaine/articles/Lascaux-Cave-and-the-rise-of-the-fake-attraction.

Martin-Sanchez, Pedro Maria, Alena Nováková, Fabiola Bastian, Claude Alabouvette & Cesareo Saiz-Jimenez. Two new species of the genus *Ochroconis*, *O. Lascauxensis* and *O. Anomala* isolated from black stains in Lascaux Cave, France. *Fungal Biology* 116, no. 5 (1 May 2012): 574–89. https://doi.org/10.1016/j.funbio.2012.02.006.

Peixotto, Becca. Perot Museum replicas of rising star caves? In-person interview. 9 July 2018.

Photos behind the scenes of the construction site at the Pont d'Arc
    Cavern. The Pont d'Arc Cavern. Accessed 21 April 2018.
    http://en.cavernedupontdarc.fr/discover-the-pont-darc-
    cavern/the-pont-d-arc-cavern-site/technological-mastery-
    to-stir-the-emotions.
Pyne, Lydia. The art of creating replicas of Ice Age cave paintings.
    Hyperallergic, 4 June 2018. https://hyperallergic.com/445675/
    the-art-of-creating-replicas-of-ice-age-cave-paintings.
Reyonlds, Natasha. Paleolithic replicas? Email interview, 27 April
    2018.
Ruiz, Aitor. Paleolithic replicas? Skype interview, 25 April 2018.
Wilkening, Susie & James Chung. 2009. *Life Stages of the Museum
    Visitor: Building Engagement over a Lifetime*. Washington, DC:
    AAM Press.

Conclusion: As Seen in the British Museum

Banksy. 2007. *Wall and Piece*. Mainaschaff: Publikat.
Banksy hoax caveman art back on display. *BBC News*, 16 May
    2018, sec. Entertainment & Arts. www.bbc.com/news/
    entertainment-arts-44140200.
Brown, Mark. Ian Hislop picks Banksy hoax for British Museum
    dissent show. *Guardian*, 16 May 2018, sec. Arts and Culture
    sec.www.theguardian.com/culture/2018/may/16/ian-hislop-
    picks-banksy-hoax-for-british-museum-dissent-show.
Cave art hoax hits British Museum, 19 May 2005. http://news.bbc.
    co.uk/2/hi/entertainment/4563751.stm.
Charney, Noah. 2015. *The Art of Forgery: The Minds, Motives and
    Methods of the Master Forgers* (1st ed.). London: Phaidon Press.
———. Is there a place for fakery in art galleries and museums?
    Aeon. Accessed 18 September 2016. https://aeon.co/essays/
    is-there-a-place-for-fakery-in-art-galleries-and-museums.
Dickens, Luke. Placing post-graffiti: the journey of the Peckham
    Rock. *Cultural Geographies; London* 15, no. 4 (October 2008):
    471–96.   http://dx.doi.org.ezproxy.lib.utexas.edu/10.1177/
    1474474008094317.
Geurds, Alexander & Laura Van Broekhoven. 2013. *Creating
    Authenticity: Authentication Processes in Ethnographic Museums*.
    Leiden: Sidestone Press.
Han, Byung-Chul. 2017. *Shanzhai: Deconstruction in Chinese*. Trans.
    by Philippa Hurd. Bilingual edition. Boston, MA: MIT Press.

Prankster infiltrates NY museums, 25 March 2005. http://news.
    bbc.co.uk/2/hi/americas/4382245.stm.
Shafir, Nir. Why fake miniatures depicting Islamic science are
    everywhere. *Aeon*. Accessed 13 September 2018. https://
    aeon.co/essays/why-fake-miniatures-depicting-islamic-
    science-are-everywhere.
Stevens, Kati. 2018. *Fake (Object Lessons)*. London: Bloomsbury
    Academic.

Acknowledgements

I couldn't have written a book like *Genuine Fakes* without the expertise and enthusiasm that so many colleagues, friends and professionals have offered over the course of this project. I owe a huge debt to: Stuart Baldwin, Nadia Berenstein, Paul Brinkman, Matthew Brown, Angela Burnley, Jill Darnell, Luke Dickens, Mike deRoos, Holly Dunsworth David Evans, Jay Ford, Mitch Fraas, Erik Goldstein, Benjamin Gross, Christopher Hallett, John Hopkins, Eliza Howlett, Bruce Hunt, Nicholas James, Lindsay Keiter, Katie Langenfeld, Rachel Lauden, Eleanor Louson, Marc Kissel, Christopher Manias, Alex McAdams, Scott McGill, Marissa Nicosia, Garrett Ozar, Linda Pellett, Becca Peixotto, Michael Press, Nick Pyenson, Megan Raby, Stephen Raby, Natasha Reynolds, Lukas Rieppel, Nicole Rudolph, Aitor Ruiz-Redondo, Josefrayn Sanchez-Perry, Clare Sauro, Christopher Schaberg, Ken Schwarz, Stephanie Strauss, Peter Tallack, Paul Taylor, Erin Thompson, Amara Thornton, Christopher Tunnell, Mike Urbancic, Gregory Urwin, Eric Williams, Audra Wolfe, Rebecca Wragg Sykes and Donna Yates.

Additionally, many institutions have been kind enough to help facilitate the research for *Genuine Fakes* through access to archives, copies of publications and materials, reprinting materials as well as interviews: Alaska Department of Fish and Game, American Museum of Natural History, Caverne du Pont d'Arc, Deutsches Museum, Eterneva, Explore.org, Harry Ransom Center at the University of Texas at Austin, the Howard Tracy Hall Foundation, *Hyperallergic*, Morgan Library & Museum, Museum of Innovation and Science, Natural History Museum (London) and Oxford Natural History Museum.

I would especially like to offer my gratitude to the
Interlibrary Loan librarians at the University of Texas at
Austin – this book simply would not have been possible
without their efforts.

My editors Jim Martin and Anna MacDiarmid have
made this project, oh, so much better through their most
welcome editorial feedback, suggestions and directions.
Krystyna Mayer's copyediting has helped sharpen and
clarify the manuscript. Holly Zemsta was kind enough to
share her thoughts on many early drafts and to truly
appreciate every bizarre factoid this project dredged up.
Over the last couple of years, my parents, Steve and Sonja
Pyne, and my sister, Molly Pyne, have offered their never-
flagging interest in this project's stories.

And, above all, I am most grateful to Stan Seibert for his
never-ending optimism that *Genuine Fakes* would become
a real book.

Index